Springer Transactions in Civil and Environmental Engineering

More information about this series at http://www.springer.com/series/13593

P.K. Swamee • B.R. Chahar

Design of Canals

Springer

P.K. Swamee
Civil Engineering
ITM University
Gurgaon, India

B.R. Chahar
Civil Engineering
Indian Institute of Technology Delhi
New Delhi, India

ISSN 2363-7633 ISSN 2363-7641 (electronic)
Springer Transactions in Civil and Environmental Engineering
ISBN 978-81-322-2321-4 ISBN 978-81-322-2322-1 (eBook)
DOI 10.1007/978-81-322-2322-1

Library of Congress Control Number: 2015936521

Springer New Delhi Heidelberg New York Dordrecht London

Printed on acid-free paper

Springer (India) Pvt. Ltd. is part of Springer Science+Business Media (www.springer.com)

Preface

Huge amounts of money are invested around the world in construction or upgradation of canals. Nearly 80–85 % of the cost of a total canal system constitutes transmission and the distribution canal networks. Due to the enormous costs involved, canal design is an area that has attracted many researchers for a long time.

The aim of this book is to provide the reader with an understanding of the analysis and design aspects of canals. The book covers the topics related to the analysis and design of water-carrying as well as sediment-transporting canals. It covers the uniform flow principles and their application in the determination of normal depth. The general principles of canal design have been covered to highlight the cost aspects and the other parameters required for the design of a canal. The other topics covered in the book relate to the determination of seepage discharge through canals of various shapes under different types of drainage positioning of drainage layers. The various topics pertain to canal design for minimum area, minimum earth work cost, minimum lining cost, minimum seepage loss, minimum evaporation loss, and their combinations. The design of contraction and expansive transitions is part of the design of cross drainage works. A chapter is also devoted to the design of these transitions.

Most of the designs are provided in a closed form that can be directly adopted by design engineers. A significant part of the book covers numerical examples. Experience has shown that complete mastery of the design project cannot be attained without familiarising oneself thoroughly with numerical procedures. For this reason, it is better not to consider numerical examples as a mere illustration but as an integral part of the general presentation.

The book is structured in a way so as to enable an engineer to design functionally efficient and least-cost canals. It is also intended to be useful for students, professional engineers, and researchers. Any suggestions for improvement of the book will be gratefully received.

Gurgaon, India — P.K. Swamee
New Delhi, India — B.R. Chahar

Contents

Authors' Biographic Sketch

Dr. P.K. Swamee is a distinguished Emeritus Professor of Civil Engineering at ITM University, Gurgaon, Haryana, India. He was formerly a Professor of Civil Engineering at the University of Roorkee (now the Indian Institute of Technology Roorkee), India. He has over fifty years of teaching, research, and industry experience in water resources engineering and has published numerous articles in international journals. Dr. Swamee is a Fellow of the Indian National Academy of Engineering.

Dr. B.R. Chahar is Professor of Civil Engineering at the Indian Institute of Technology Delhi, India. He was Visiting Professor at the Ecole des mines de Saint-Etienne, France, and Asian Institute of Technology, Bangkok. Dr. Chahar received his Ph.D. in Civil Engineering from IIT Roorkee and M.Tech. in Civil Engineering with specialisation in Water Resources Engineering from IIT Kharagpur. He has 22 years of teaching, research and industry experience in water resources engineering and has published numerous articles in reputed journals and proceedings. Dr. Chahar is a recipient of several scholarships and awards, including Visiting International Fellowship of EWRI of ASCE, Young Scientist of DST, and AICTE Career Award for Young Teachers. He is Fellow of ISH, IWWA, IE(I), and IWRS and life member of ISTE, IAH, and ASCE.

Chapter 1
Introduction

Abstract Irrigation has been practiced since the beginning of civilization. A canal is used to convey water from a source to a destination for irrigation, industrial, or domestic use. The canal must be capable of transporting water between the source and the destination in a reliable and cost-effective manner. A brief history of developments in canals is presented. An outline of all chapters and scope of the book are also reported in the chapter.

Keywords Irrigation • Canal • Uniform flow • Resistance equation • Seepage loss • Canal lining • Canal alignment • Section cost

1.1 General

A canal is used to convey water from a source to a destination for irrigation, industrial or domestic use. The canal must be capable of transporting water between the source and the destination in a safe and cost-effective manner.

Irrigation has been practiced since the beginning of civilization. In due course of time, it was discovered that water could be diverted from a stream and conveyed under gravity flow in a ditch to a lower land. The large-scale diversion of water for irrigation was established in historic times in Mesopotamia 5,000–6,000 years ago (Butler 1960). The Sumerians developed extensive and intricate networks of irrigation channels around 3,000 B.C. Some of the canals still exist in usable form, e.g., Shatt-al Hai canal in Iraq built around the middle of the third millennium B.C. (Lewis 1960). Historical developments in irrigation canals are reported by Framji et al. (1982) for the period starting from the old civilizations in India, China, Egypt, etc., to the post-Christ period in all over the world.

During the latter half of nineteenth century, investigators worked towards the rational design of irrigation canals. Kennedy (1895), Lindley (1919), Lacey (1930, 1940, 1946), and others developed the regime equations for design of stable channels. The stable channel design by limiting tractive force approach was originated by Lane (1937, 1955) and others. Framji (1972) has outlined design methods for irrigation canals practiced in different countries.

P.K. Swamee, B.R. Chahar, *Design of Canals*, Springer Transactions in Civil and Environmental Engineering, DOI 10.1007/978-81-322-2322-1_1

As the slope required for irrigation canals is much less than the ground slope, falls are provided to save the earthwork in filling cost. These falls are utilized in power generation. On the other hand, canals are exclusively constructed for power generation. Run-of-the-river hydropower channel is one such example in which very little storage is required. In both run-of-the-river and storage schemes of hydropower generation, water is diverted from one point on the river to the turbines and then from turbine house back to the river.

Ship canals or enlarged barge canals are constructed to accommodate ships. As ship canals are not supposed to carry any discharge, the dimensions for ship canals are largely governed by the size of the largest ships the canal is supposed to carry. Ship canals may be constructed to create a shortcut for avoiding lengthy detours, to create a shipping route between two land-locked seas or lakes, to provide inland cities with a direct shipping link to the sea, etc.

1.2 Objective Function

The optimal design of a canal consists of minimization of an objective function which is subjected to certain constraints. The known parameters are flow discharge, longitudinal bed slope of canal, and the canal surface roughness. There are various objective functions such as flow area, earthwork cost, lining cost, seepage loss, evaporation loss, and their combinations (Swamee et al. 2000a, b, c, 2001a, b, 2002a, b; Chahar 2000; Basu 2013). Apart from costs, reliability is another important objective of the canal design. However, there has been no attempt in this direction.

As artificial channels have objective function, natural channels also have objectives. A natural channel is a stream in equilibrium, which is neither silting nor scouring over a period of time. Obviously, such a stream has developed a cross-sectional area of flow through natural processes of deposition and scour. Using Lacey's equations for stable channel geometry, and using geometric programming, Swamee (2000) synthesized an objective function for stable alluvial channels. On the other hand, in a similar manner, Swamee et al. (2008) found an objective function for river Brahmaputra. Chapter 2 formulates objective functions.

1.3 Uniform Flow

Canals are designed for uniform flow considering economy and reliability. Uniform flow is described by a resistance equation. Whereas the economy is achieved by minimization of cost, maximum reliability is realized by delivery of discharge with the least frequency of failures.

The first uniform flow equation was derived by Antoine de Chézy (Rouse and Ince 1963) in 1769 mathematically from the definition of uniform flow with the assumption that the force resisting the flow per unit area of the stream bed is proportional to the square of the velocity. Other commonly used equations are due to Ganguillet and Kutter and Manning. Though the Ganguillet and Kutter equation appears cumbersome, it became the most popular equation of its time, and many tables and charts (Garrett 1948) were prepared for its use. The popularly known Manning's equation was proposed by Gauckler (Williams 1970; Rouse 1956) in 1867 and Hagen in 1876; Hagen obtained the same relationship using Upper Ganga Canal data, which were collected by Cunningham with observations at Roorkee (Mital 1986). Later, Vallot in 1887 and Thrupp in 1888 (Williams 1970) also obtained the same equation. For high roughness projections and for high enough Reynolds numbers, friction is independent of Reynolds number and depends only on the hydraulic radius of the flow for a given type of channel surface, and hence, Manning's equation is applicable only to the fully rough turbulent flow and in a limited bandwidth of relative roughness (Christensen 1984). For other flow conditions, the Colebrook equation is more appropriate than the Manning's equation ("Friction" 1963). A more general resistance equation based on roughness height was given by Swamee (1994). Chapter 3 describes uniform flow equations for viscous flow, turbulent flow, and sediment-transporting channels.

1.4 General Principles

In designing canals, various factors are considered, for example, the kind of material forming the channel surface which determines the roughness coefficient, the minimum permissible velocity to avoid deposition of silt or debris, the limiting velocity to avoid erosion of the channel surface, and the topography of the channel route which fixes the channel bed slope. Chapter 4 describes general principles of canal design, which includes essential input parameters and safety and system constraints.

1.5 Minimum Area Section

A canal based on minimization of flow area objective function is also a maximum velocity canal. Several literature is available on minimum area section (Chahar 2005, 2007), but majority of literature used Manning's equation while only few (Chahar 2000; Swamee 1995) have used general resistance equation in design of such sections. Design of minimum flow area section has been covered in Chap. 5.

1.6 Minimum Cost Canal Section

Design of a minimum cost canal section involves minimization of the sum of earthwork cost and cost of lining subject to uniform flow condition in the canal (Swamee et al. 2000b, 2001a). In general, the cost of earthwork varies with canal depth. Chapter 6 discusses the optimization method to obtain minimum cost canal sections.

1.7 Minimum Water Loss Section

An irrigation canal may be an unlined canal or a lined canal. The loss of water due to seepage and evaporation from canals constitutes a substantial part of the available water. By the time the water reaches the field, up to one-half of the water supplied at the head of the canal may be lost in transit (Sharma and Chawla 1975). The evaporation loss is important particularly for a long channel carrying a small discharge in water-scarce areas of arid regions. The seepage loss from canals depends on hydraulic conductivity of the subsoils, canal geometry, and location of water table relative to the canal. The seepage loss results not only in depleted fresh water resources but also causes water logging, salinization, groundwater contamination, and health hazards. Canal lining checks the seepage from a canal. To minimize seepage and to transport water efficiently, lined canals were envisaged. Nowadays, a new canal, preferably, is constructed as a lined canal. A perfect lining would prevent all the seepage loss, but canal lining deteriorates with time. An examination of canals by Wachyan and Rushton (1987) indicates that even with the greatest care, the lining does not remain perfect. A well-maintained canal with 99 % perfect lining reduces seepage about 30–40 % (Wachyan and Rushton 1987). Cracks in canal lining develop due to settlement of the subgrade, weed growth in the canal, construction defects, use of inferior-quality lining materials, weathering, etc. The thickness of the lining material is small and cracks may develop anywhere on the perimeter. Therefore, seepage from a canal with cracked lining is likely to approach the quantity of seepage from an unlined canal. Chapter 7 deals with the design of canal sections considering seepage and evaporation water losses as reported by Swamee et al. (2000a, 2001b, 2002a).

1.8 Minimum Overall Cost Section

Design of a minimum cost canal section involves minimization of the sum of earthwork cost, cost of lining, and cost of water lost as seepage and evaporation subject to uniform flow condition in the canal. Swamee et al. (2000c) took into account all such costs in the design of minimum cost canal sections. The optimal canal design equations considering overall cost due to lining, earthwork, and water loss are highlighted in Chap. 8.

1.9 Canal Transitions

A canal from source to destination may be of several hundred kilometers. The discharge in the canal varies along the length due to diversion and losses; therefore, a reduced canal section matching with the discharge is adopted. Canal section may also change at flumes, siphons, and aqueducts. A canal transition involving an expansion or contraction of the section is required whenever there is change in the canal section. A transition is a structure of short length; thus, the cost aspect of transitions is not considered in their design. Chapter 9 gives design procedure for both contraction and expansion transitions.

1.10 Transmission Canal

A transmission canal conveys water from the source to a distribution canal. Many times, the area to be irrigated lies very far from the source, requiring long transmission canals. Though there is no withdrawal from a transmission canal, it loses water on account of seepage and evaporation. Hence, it is not economical to continue the same section throughout the length of a long transmission canal. Instead a transmission canal should be divided into subsections or reaches, and the cross section for each of the subsections must be designed separately (Swamee et al. 2002b). This would result in reduced cross sections in the subsequent reaches. The reduced cross section not only results in cost saving for earthwork, lining, and water lost but also requires less cost in land acquisition, construction of bridges, and cross-drainage works. Chapter 10 addresses the problem of design of transmission canal.

1.11 Canal Route Alignment

The total cost of a canal project depends upon the alignment. The alignment is the feasible path or route from a source location to the desired destination. A canal has to be aligned in such a way that it covers the entire area proposed to be irrigated with the shortest possible length, and at the same time, its cost including the cost of cross-drainage works is a minimum. A shorter length of canal ensures less loss of head due to friction and smaller loss of discharge due to seepage and evaporation, so that additional areas can be brought under cultivation. Canal alignment may be contour canal, side slope canal, or ridge canal as per terrain of command area. A contour canal irrigates only one side of the canal and it crosses a number of valleys; thus, it involves different types of cross-drainage works such as aqueducts, under tunnels, super passages, etc. A side slope canal is aligned at right angles to the contours of a country. A watershed or ridge canal irrigates the areas on both sides. Cross-drainage works are completely eliminated in watershed and side slope canals. The main canal is generally carried on a contour alignment, until either it commands the full area to be irrigated or it attains the top of a watershed to become a watershed canal thereafter. Branch canals and distributaries take off from a canal from or near the points where the canal crosses the watershed. The alignment of a canal is decided after a careful consideration of the economy. Several alignments between the source and the destination may be possible. An alignment mainly depends on the topography. Out of many alignments, few may not be feasible to construct due to construction-related problems. Canals are aligned as far as possible in partial cutting and partial filling. Deep cutting or high embankments are generally avoided by suitable detouring after comparing the overall costs of the alternative alignments. Land cost varies with land use pattern, resettlement, and rehabilitation cost, environmental cost, and alignment of the canal; the cost of canal falls/drops/cross-drainage works varies with the type and size of structure. The maximization of economy is achievable by minimization of the total cost of canal route alignment considering all possible cost factors (Basu 2013; Swamee and Chahar 2013). This type of canal alignment problem is solved in Chap. 11.

1.12 Mathematical Terms

- Adjoint variables: The undetermined multipliers used in formation of Hamiltonian are called the adjoint variables. In the present case, these variables are functions of space.
- Design variables: These are the variables that are determined in a design process.
- Grid search method: This is a method of optimization in which the values of objective function $F(x_1, x_2, \ldots x_n)$ are computed at equispaced points in the domain of variables $x_1, x_2, \ldots x_n$, and the optimum is found by comparing these values. The process is repeated in the neighborhood of the located optima till desired accuracy is achieved. Thus, it is a brute force method of optimization.
- Hamiltonian: This is a function formed by adding the integrand of the objective functional with the state equations through undetermined multipliers.
- Objective function: It is a function expressing the requirements of an engineering system like cost, efficiency, and power consumption. When this function is optimized (minimized or maximized), it yields the best engineering design.
- Optimal control problem: In this problem the objective function is in the form of an integral (called functional) involving the design variable as an unknown function of space along with state variables. The restrictions imposed on state variables and the design variables are in the form of differential equations. The original problem as formulated in electrical engineering is in time domain. In this paper, it is converted in space domain.

- Penalty function: The objective function is penalized by this function if it departs from the restriction placed on a design variable or control variable. The restriction placed on the design variable b is $b(L) = b_L$, and the restriction placed on the control variable w is $w(0) = w(L) = 0$.

1.13 Scope of the Book

The book is structured in such a way that it enables engineers to design least-cost canals. It is intended that students, professional engineers, and researchers will benefit from the canal design topics covered in this book. Hopefully, it will turn out to be a reference book to irrigation engineers.

References

Basu S (2013) Optimal alignment of a canal route. PhD thesis, Indian Institute of Technology Delhi, New Delhi

Butler SS (1960) Irrigation systems of the Tigris and Euphrates valleys. J Irrig Drain Eng ASCE 86(4):59–79

Chahar BR (2000) Optimal design of channel sections considering seepage and evaporation losses. Ph.D. thesis, Department of Civil Engineering, University of Roorkee, Roorkee, India

Chahar BR (2005) Optimal design of a parabolic channel section. J Irrig Drain Eng ASCE 131(6):546–554

Chahar BR (2007) Optimal design of special class of curvilinear bottomed channel section. J Hydraul Eng ASCE 133(5):451–460

Christensen BA (1984) Discussion on 'Flow velocities in pipelines', by Richard R. Pomeroy. J Hydraul Eng ASCE 10(10):1510–1512

Framji KK (1972) Design practices of irrigation canals in the world. International Commission on Irrigation and Drainage, New Delhi

Framji KK, Garg BC, Luthra SDL (1982) Irrigation and drainage in the world: a global review. International Commission on Irrigation and Drainage, New Delhi

Friction factors in open channels (1963) Progress report of the task force on friction in open channels of the Committee on Hydromechanics of the Hydraulics Division. J Hydraul Eng Div ASCE 89(2):97–137

Garrett AFF (1948) Hydraulic diagrams for the design of channels in earth by Kutter's formula. Central Board of Irrigation, Kennedy House, Shimla

Kennedy RG (1895) The prevention of silting in irrigation canals. Proc Inst Civil Eng Lond 119:281–290

Lacey G (1930) Sable channel in alluvium. Proc Inst Civil Eng Lond 229:259–285

Lacey G (1940) Regime flow in incoherent alluvium, Publication no. 20. Central Board of Irrigation, Kennedy House, Shimla

Lacey G (1946) A general theory of flow in alluvium. J Inst Civil Eng Lond 27:16–47

Lane EW (1937) Stable channels in erodible materials. Trans ASCE 102:123–142

Lane EW (1955) Design of stable channels. Trans ASCE 120:1234–1260

Lewis MR (1960) Heritage of irrigation in Iraq. J Irrig Drain Eng ASCE 86(2):9–19

Lindley ES (1919) Regime channels. In: Proceedings of Punjab Engineering Congress, Punjab Engineering Society, Lahore, India (now Pakistan), 7:63–74

Mital KV (1986) History of Thomason College of Engineering. University of Roorkee, Roorkee
Rouse H (1956) Discussion on 'A note on the Manning formula' by Ven Te Chow. Trans Am Geophys Union 37(3):327–328
Rouse H, Ince S (1963) History of hydraulics. Dover Publications Inc., New York
Sharma HD, Chawla AS (1975) Manual of canal lining, Technical report no. 14. Central Board of Irrigation and Power, New Delhi
Swamee PK (1994) Normal depth equations for irrigation canals. J Irrig Drain Eng ASCE 120(5):942–948
Swamee PK (1995) Optimal irrigation canal sections. J Irrig Drain Eng ASCE 121(6):467–469
Swamee PK (2000) Stable channel objective function. Int J Sediment Res 15(4):434–439
Swamee PK, Mishra GC, Chahar BR (2000a) Design of minimum seepage loss canal sections. J Irrig Drain Eng ASCE 126(1):28–32
Swamee PK, Mishra GC, Chahar BR (2000b) Minimum cost design of lined canal sections. Water Resour Manag 14(1):1–12
Swamee PK, Mishra GC, Chahar BR (2000c) Comprehensive design of minimum cost irrigation canal sections. J Irrig Drain Eng ASCE 126(5):322–327
Swamee PK, Mishra GC, Chahar BR (2001a) Design of minimum earthwork cost sections. Water Resour Manag 15(1):17–30
Swamee PK, Mishra GC, Chahar BR (2001b) Design of minimum seepage loss canal sections with drainage layer at shallow depth. J Irrig Drain Eng ASCE 127(5):287–294
Swamee PK, Mishra GC, Chahar BR (2002a) Design of minimum water loss canal sections. J Hydraul Res 40(2):215–220
Swamee PK, Mishra GC, Chahar BR (2002b) Optimal design of transmission canal. J Irrig Drain Eng ASCE 128(4):234–243
Swamee PK, Chahar BR (2013) Optimal alignment of a canal route. Proc ICE Water Manag 166(WM8):422–431
Swamee PK, Sharma N, Dwivedi A (2008) Lacey regime equations for River Brahmaputra. J Hydraul Res IAHR 46(5):707–710
Wachyan E, Rushton KR (1987) Water losses from irrigation canals. J Hydrol 92(3–4):275–288
Williams GP (1970) Manning formula – a misnomer? J Hydraul Eng ASCE 96(1):193–201

Chapter 2
Objective Functions

Abstract The optimal design of a canal consists of minimization of an objective function which is subjected to certain constraints. The known parameters are flow discharge, longitudinal bed slope of canal, and the canal surface roughness. There are various objective functions such as flow area, earthwork cost, lining cost, seepage loss, evaporation loss, and their combinations. This chapter describes geometric properties and seepage loss functions of commonly used channel sections as well as computation of lining cost, earthwork cost, cost of water lost as seepage and evaporation loss, and capitalized cost. A unification of all these costs results in cost function of rigid boundary canals. A natural channel is a stream in equilibrium, which is neither silting nor scouring over a period of time. Such a stable channel develops a cross-sectional area of flow through natural processes of deposition and scour. Using Lacey's equations for stable channel geometry and using geometric programming, an objective function for stable alluvial channels can be synthesized. Thus, this chapter formulates objective functions for rigid boundary canals and mobile boundary (natural) canals.

Keywords Natural channel • Stable channel • Cost function • Objective function • Annuity • Capitalization • Seepage loss • Evaporation loss • Geometric properties • Lining cost • Earthwork cost • Cost of water lost

A design problem, like a canal design, is an optimization problem where an objective function is minimized subject to various constraints. In water resource projects, this function is cost-benefit ratio. Here, as benefits are constant, it is sufficient to minimize cost. The cost consists of capital costs and recurring costs. Costs of land, excavation, and lining are capital costs, whereas cost maintenance and water loss on account of seepage and evaporation are recurring costs. In this book, the cost of land is not considered as it involves alignment problem involving topography. Thus, the design of canal section is considered with known entities being flow discharge, longitudinal canal bed slope, and the canal surface roughness. As the general problem is an involved one, sometimes it is simplified by considering the flow area being the objective function. Such a canal is a minimum area or maximum velocity canal or best hydraulic section. The best hydraulic section has the minimum flow area and flow perimeter for a given discharge but not necessarily

P.K. Swamee, B.R. Chahar, *Design of Canals*, Springer Transactions in Civil and Environmental Engineering, DOI 10.1007/978-81-322-2322-1_2

the most economical section. A network of canals represents a major cost item in an irrigation project, and the economy of the canal network is vital. The maximum economy is achieved by minimizing the cost of the canals. The design of minimum cost irrigation canals involves minimization of the sum of earthwork cost which varies with canal depth, cost of lining, and cost of water lost as seepage and evaporation subject to uniform flow condition in the canal. Such minimum cost canal design problem results in nonlinear objective function and nonlinear equality constraint, making the problem hard to solve analytically. In the book, the earthwork and the lining costs have been considered for the flow section only.

2.1 Flow Area

Figure 2.1 depicts commonly used canal sections. Triangular sections are generally constructed for carrying small discharges. For a triangular section (Fig. 2.1a), the flow area A is given by

$$A = my_n^2 \tag{2.1.1}$$

where m = side slope and y_n = normal depth (m). Rectangular sections are generally constructed for moderate discharges. For a rectangular section of bed width b (m) (see Fig. 2.1b), the flow area is

$$A = by_n \tag{2.1.2}$$

For carrying large discharges, rectangular sections are not preferred. This is on account of stability of side slopes. Vertical side walls require large thickness to resist the earth pressure. On the other hand, sloping side walls require less thickness. For a trapezoidal section of bed width b and side slope m (see Fig. 2.1c), the flow area is

$$A = y_n\left(b + my_n\right) \tag{2.1.3}$$

For small discharges, semicircular sections are often adopted for irrigation canals. For a circular section of diameter D (m) (see Fig. 2.1d), the flow area is

$$A = 0.25D^2\left[\cos^{-1}\left(1 - 2\eta_n\right) - 2\left(1 - 2\eta_n\right)\sqrt{\eta_n\left(1 - \eta_n\right)}\right] \tag{2.1.4}$$

where $\eta_n = y_n/D$. A power law section is described by

$$Y = \left|k_p X\right|^p \tag{2.1.5}$$

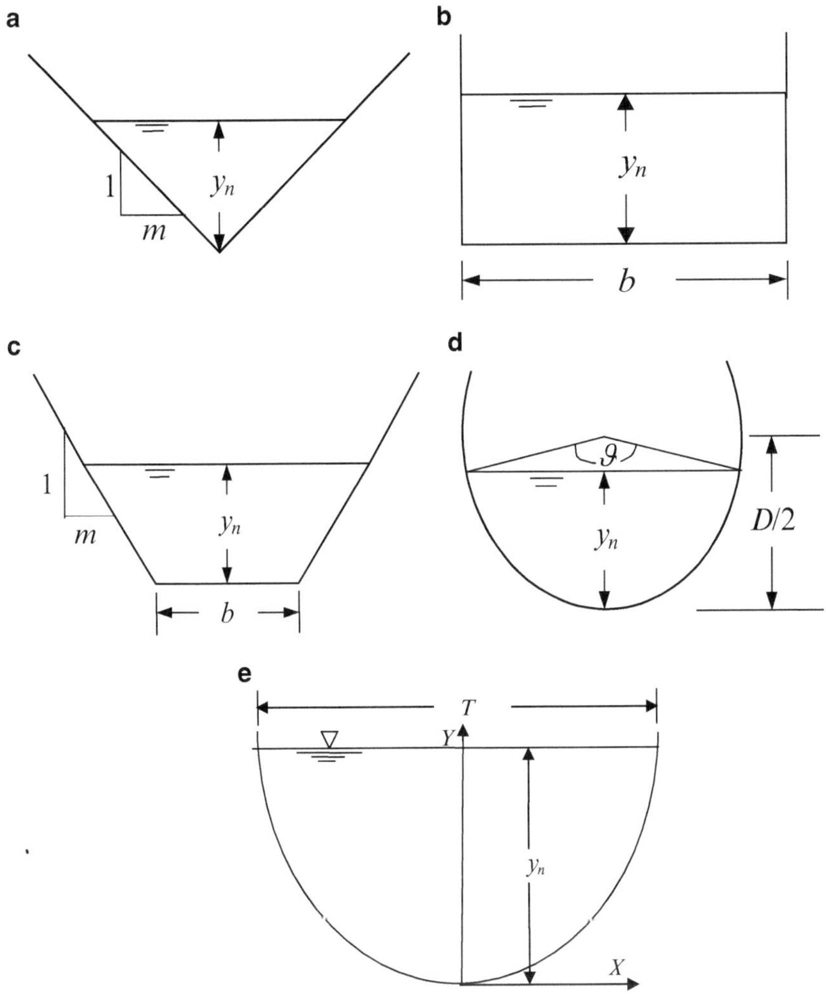

Fig. 2.1 Canal sections: (**a**) triangular, (**b**) rectangular, (**c**) trapezoidal, (**d**) circular, and (**e**) parabolic

where X and Y = horizontal and vertical coordinates, respectively, as shown in Fig. 2.1e, k_p = coefficient, and p = exponent. For $p = 1$, the exponential section is a triangle of side slope $m = 1/k_p$; for $p = 2$, it is a parabola of latus rectum $1/k_p^2$. For $p = \infty$, $Y = 0$ when $|k_p X < 1|$ and $Y = \infty$ when $|k_p X|$ just exceeds unity. Thus, a rectangle of bed width $2/k_p$ is obtained. The width of water surface for the power law section is given by

$$T = \frac{2}{k_p} y_n^{1/p} \tag{2.1.6}$$

The area of the section for depth y_n is given by

$$A = \frac{2py_n^{(p+1)/p}}{k_p(p+1)} \tag{2.1.7}$$

For $p = 2$ in Eq. (2.1.5), the power law section becomes a parabola having the equation

$$Y = k_p^2 X^2 \tag{2.1.8}$$

This parabola has latus rectum $a = 1/k_p^2$. The area of this section is given by

$$A = \frac{4y_n^{3/2}}{3k_p} = \frac{4}{3} y_n \sqrt{a y_n} \tag{2.1.9}$$

For $p = 0.5$ in Eq. (2.1.5), the power law section becomes an inverse parabola having the equation

$$Y = \sqrt{k_p X} \tag{2.1.10}$$

This parabola has latus rectum $a = k_p$. The area of this section is given by

$$A = \frac{2y_n^3}{3k_p} = \frac{2y_n^3}{3a} \tag{2.1.11}$$

On the other hand, for $p = \infty$, the section gets converted into a rectangle of bed width $2/k_p$. The area of this section is $2y_n/k_p$.

2.2 Lining Cost

The lining cost depends on the extent of canal surface area to be lined and the type of lining material to be used. Considering unit cost of lining (cost per unit surface area covered) to be independent of depth of placement, the cost of lining C_L (monetary unit per unit length of canal, e.g., ₹/m) is expressed as

$$C_L = c_L P \tag{2.2.1}$$

where c_L = unit cost of lining (monetary unit per unit area, e.g., ₹/m^2) and P = channel perimeter (m). Table 2.1 lists the perimeter of various shapes of a canal section. Once the perimeter is known, the following are the lining costs for various sections:

Table 2.1 Geometrical properties of canal sections

Section shape	Geometric elements		
	Flow perimeter	Area of flow	Depth of centroid of area
	P	A	$\overline{y}$
(1)	(2)	(3)	(4)
Triangular	$2y_n\sqrt{1+m^2}$	my_n^2	$y_n/3$
Rectangular	$b+2y_n$	by_n	$y_n/2$
Trapezoidal	$b+2y_n\sqrt{1+m^2}$	$(b+my_n)\,y_n$	$\frac{y_n}{6}\left(\frac{3b+2my_n}{b+my_n}\right)$
Circular	$0.5D\vartheta$	$\frac{D^2}{8}(\vartheta-\sin\vartheta)$	$\frac{D}{6}\left(\frac{6\sin\frac{\vartheta}{2}-3\cos\frac{\vartheta}{2}-2\sin^3\frac{\vartheta}{2}}{\vartheta-\sin\vartheta}\right)$
Parabolic	$\frac{a}{2}\Big[2\sqrt{\eta_n}\,(1+4\eta_n) + \ln\big(2\sqrt{\eta_n} + \sqrt{1+4\eta_n}\big)\Big]$	$\frac{4}{3}y_n\sqrt{ay_n}$	$\frac{2}{5}y_n$

where $\vartheta = 2\cos^{-1}(1-2\eta_n)$, wherein $\eta_n = y_n/D$ for circle and $\eta_n = y_n/a$ for parabola

$$C_L = 2c_L y_n\sqrt{1+m^2} \quad \text{triangular section} \tag{2.2.2}$$

$$C_L = c_L\,(b+2y_n) \quad \text{rectangular section} \tag{2.2.3}$$

$$C_L = c_L\left(b+2y_n\sqrt{1+m^2}\right) \quad \text{trapezoidal section} \tag{2.2.4}$$

$$C_L = c_L D\cos^{-1}(1-2\eta_n) \quad \text{circular section} \tag{2.2.5}$$

$$C_L = \frac{c_L a}{2}\left[2\sqrt{\eta_n}\,(1+4\eta_n) + \ln\left(2\sqrt{\eta_n}+\sqrt{1+4\eta_n}\right)\right] \quad \text{parabolic section} \tag{2.2.6}$$

where $\eta_n = y_n/D$ for circular section and y_n/a for parabolic section.

2.3 Earthwork Cost

Earthwork in the form of cutting and/or filling along the canal alignment is required for providing canal flow area. Earthwork cost is the major cost item for a canal passing through hard/firm strata, where lining may not be required. Sometimes canals are lined with low-cost lining materials, in which case the cost of the earthwork is more significant than the cost of lining. The cost of earthwork depends on the volume and depth of cut and fill. It also depends on the strata to be excavated and the distance of haulage if required in transporting the soil materials. The cost

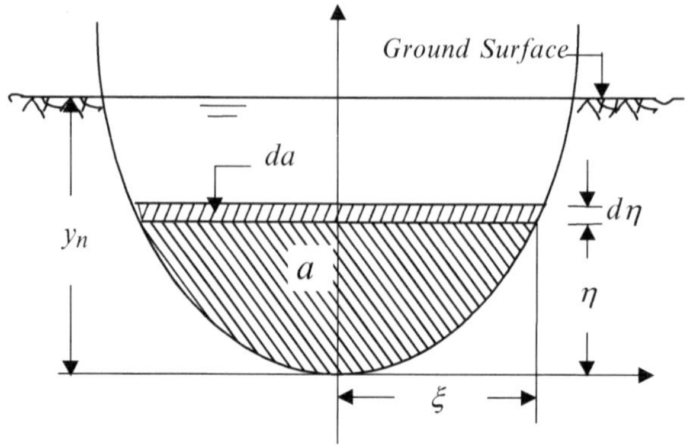

Fig. 2.2 Definition sketch for earthwork cost

function consists of earthwork cost of unit length of the canal. For a canal section with the normal water surface at the average ground level as shown in Fig. 2.2, the earthwork cost C_e (monetary unit per unit length of canal, e.g., ₹/m) is given by

$$C_e = c_e A + c_r \int_{\text{Area}} (y_n - \eta)\, da = c_e A + c_r \int_0^{y_n} (y_n - \eta)\, 2\xi \; d\eta \qquad (2.3.1)$$

where c_e = cost per unit volume of earthwork at ground level (₹/m^3), c_r = the additional cost per unit volume of excavation per unit depth (₹/m^4), η = vertical ordinate, ξ = half the width of excavation at the ordinate η, $d\eta$ = incremental vertical ordinate, a = excavation area up to height η, and da = incremental excavation area. It was assumed in Eq. (2.3.1) that the cost per unit volume of excavation is linear function of the depth of excavation. Integrating Eq. (2.3.1) by parts resulted in

$$C_e = c_e A + c_r |a\,(y_n - \eta)|_0^{y_n} + c_r \int_0^{y_n} a d\eta = c_e A + c_r A \overline{y} \qquad (2.3.2)$$

where $\overline{y}$ = depth (m) of the centroid of the area of excavation from the ground surface. Table 2.1 lists A and $\overline{y}$ for triangular, rectangular, trapezoidal, circular, and parabolic canal sections.

2.4 Annual Water Loss Cost

Annual cost of water loss per unit length of canal consists of costs of annual seepage loss and annual evaporation loss from one meter length of the canal. In Sects. 2.4.1 and 2.4.2 given below, the water loss per second is described.

2.4.1 Seepage Loss

The seepage loss from irrigation canals constitutes a substantial part of usable water. According to the Indian Bureau of Standards (1980), the loss of water by seepage from unlined canals in India generally varies from 0.3 to 7.0 m^3/s per 10^6 m^2 of wetted surface. Providing perfect lining can prevent seepage loss from canals, but cracks in lining develop due to several reasons and performance of canal lining deteriorates with time. An examination of canals by Wachyan and Rushton (1987) indicated that even with the greatest care, the lining does not remain perfect. A well-maintained canal with 99 % perfect lining reduces seepage about 30–40 % only (Wachyan and Rushton 1987). Thus, significant seepage losses occur from a canal even if it is lined. The seepage loss from canals is primarily governed by hydraulic conductivity of the subsoil, canal geometry, and potential difference between the canal and the aquifer underneath which in turn depends on the initial and boundary conditions. Seepage losses are also influenced by clogging of the canal surfaces depending on the suspended sediment content of the water and on the grain-size distribution of the suspended sediment particles. The clogging process can decrease the seepage discharge both through bottom and slopes. Thus, the seepage loss can change within time, and under certain conditions, it can diminish. Therefore, the seepage loss can be higher at the beginning of the canal operation and can be lower after a few years of operation.

The steady seepage loss from an unlined or a cracked lined canal complies with Darcy's law. Swamee et al. (2000) expressed the analytic solutions of seepage loss in the following simple form:

$$q_s = k y_n F_s \tag{2.4.1}$$

where q_s = seepage loss per unit length of canal (m^2/s), k = hydraulic conductivity of porous medium (m/s), and F_s = seepage function, which is a function of channel geometry and boundary condition. Depending upon the geometry of the flow domain, one of the following boundary conditions may exist: (i) porous medium underlain by an impermeable layer at a finite depth; (ii) porous medium underlain by a drainage layer at a finite depth, and water table is above the top of the drainage layer; (iii) porous medium underlain by a drainage layer at finite depth, and water table is below the top of the drainage layer; (iv) water table at a finite depth in a porous medium of infinite depth; and (v) porous medium of infinite depth in which water table is at infinite depth or a drainage layer and water table are both at infinite depth. Case (iii) and case (v) are important in canal design considering seepage.

Porous medium having infinite depth: Analytic solutions for the seepage function for case (v) are available for known canal dimensions (Harr 1962; Morel-Seytoux 1964; Polubarinova-Kochina 1962). The analytical form of these solutions, which contain improper integrals and unknown implicit state variables, is not convenient in estimating seepage from the existing canals and in designing canals considering seepage loss. These methods have been simplified using numerical methods for easy

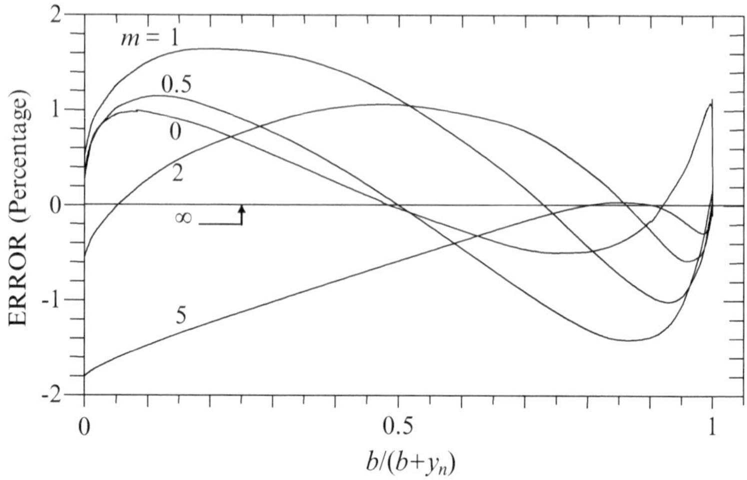

Fig. 2.3 Error diagram – Eq. (2.4.4)

computation of seepage function by Swamee et al. (2000). The simplified equations are as follows:

$$F_s = \left\{[\pi(4-\pi)]^{1.3} + (2m)^{1.3}\right\}^{0.77} \quad \text{triangular section} \tag{2.4.2}$$

$$F_s = \left\{[\pi(4-\pi)]^{0.77} + \left(\frac{b}{y_n}\right)^{0.77}\right\}^{1.3} \quad \text{rectangular section} \tag{2.4.3}$$

$$F_s = \left(\left\{[\pi(4-\pi)]^{1.3} + (2m)^{1.3}\right\}^{\frac{0.77+0.462m}{1.3+0.6m}} + \left(\frac{b}{y_n}\right)^{\frac{1+0.6m}{1.3+0.6m}}\right)^{\frac{1.3+0.6m}{1+0.6m}}$$

trapezoidal section (2.4.4)

Figure 2.3 depicts the errors involved in Eq. (2.4.4) for computation of seepage from a trapezoidal canal. A perusal of Fig. 2.3 shows the maximum error as 1.8 % for the triangular section ($b = 0$). For the rectangular section ($m = 0$), the maximum error is within 1 %. The error in the practical range is less than 0.9 % for the triangular section ($0.5 \le m \le 2.5$), 0.5 % for the rectangular section ($0.5 \le b/y_n \le 10$), and 1.4 % for the trapezoidal section ($0.5 \le m \le 5$ and $0.5 \le b/y_n \le 10$).

A very narrow and deep channel is called a *slit*. Solution for seepage loss from a slit forms a particular case of the solutions given for triangular, rectangular, and trapezoidal canals. For a slit, the width T at water surface approaches zero, i.e., the ratio $T/y_n \to 0$, which means $m \to 0$ for triangular section, $b/y_n \to 0$ for rectangular channel, and both $m \to 0$ and $b/y_n \to 0$ for a trapezoidal canal. With these conditions Eqs. (2.4.2), (2.4.3), and (2.4.4) give seepage function for a slit as

$$F_s = \pi\,(4 - \pi) \tag{2.4.5}$$

which is near exact as the error is within 0.12 %, since the exact seepage function for a slit (Chahar 2000; Swamee et al. 2001b) is

$$F_s = \pi^2/(4G) \tag{2.4.6}$$

where $G =$ Catalan's constant. A *strip* is a reverse case of a slit, i.e., for a strip $b/y_n \to \infty$. This happens for a very wide and shallow channel. The seepage function for a strip channel is given by (Swamee et al. 2001b)

$$F_s = \frac{b}{y_n} + \frac{16G}{\pi^2} \tag{2.4.7}$$

Since b/y_n is very large in comparison to $16G/\pi^2$, Eq. (2.4.7) reduces to

$$F_s = b/y_n \tag{2.4.8}$$

Solution with limit $m \to \infty$ in Eq. (2.4.2) or $b/y \to \infty$ in Eq. (2.4.3) gives the seepage function for a strip as given by Eq. (2.4.8).

The analytic solutions for seepage discharge from canals pertain to triangular, rectangular, and trapezoidal sections which are polygonal sections. Apart from the usual assumptions of homogeneity and isotropy of the conducting porous medium, these solutions are mostly based on the assumption of an infinite depth drainage layer. At infinite depth, the streamlines become vertical; hence, the seepage width attains its potential (maximum) value, and seepage occurs under unit hydraulic gradient.

For a circular section, Swamee and Kashyap (2001) gave the following equation for seepage function:

$$F_s = \eta_n^{-1}\left\{\left[2\sqrt{\left(\eta_n^{-1} - 1\right)} + 6.24\eta_n^{-1}\left(\eta_n^{-1} - 1\right)^{-1.65}\right]^{-0.5} + 0.584\left[\eta_n^{-1} + 3.55\eta_n^{-1}\left(\eta_n^{-1} - 1\right)^{0.8}\right]^{-0.5}\right\}^{-2} \tag{2.4.9}$$

where $\eta_n = y_n/D$. For power law section with $p \geq 1$, the following equation is obtained (Swamee and Kashyap 2001):

$$F_s = \left\{ [\pi (4-\pi)]^{\frac{0.7p+0.3}{0.91p-0.14}} + \left(\frac{2}{k_p y_n^{(p-1)/p}} \right)^{\frac{0.7p+0.3}{0.91p-0.14}} \right\}^{\frac{0.91p-0.14}{0.7p+0.3}} \tag{2.4.10}$$

Similarly, for the exponent range $0.25 \leq p \leq 0.75$, the following equation for F_s was obtained:

$$F_s = \frac{2}{k_p} y_n^{(1-p)/p} \left\{ \left(1 + \frac{250}{1+0.8p} \phi^{-\frac{1}{1+8p}} \right)^{-\frac{2.5}{p}} + \left[1 + (0.1 + 1.6p)\, \phi^{-0.25p^{-2}} \right]^{-\frac{2.5}{p}} \right\}^{-0.4p} \tag{2.4.11}$$

where $\phi = y_n / k_p^{p/(1-p)}$. It can be seen that for $p = 1$, Eq. (2.4.10) is converted to

$$F_s = \left\{ [\pi (4-\pi)]^{1.3} + \left(\frac{2}{k_p} \right)^{1.3} \right\}^{0.77} \tag{2.4.12}$$

which is the same as Eq. (2.4.2). On the other hand, for $p = \infty$, Eq. (2.4.10) is changed to

$$F_s = \left\{ [\pi (4-\pi)]^{0.77} + \left(\frac{2}{k_p y_n} \right)^{0.77} \right\}^{1.3} \tag{2.4.13}$$

which is the same as Eq. (2.4.3). Setting $p = 2$ in Eq. (2.4.10), the parabolic section yielded the following equation:

$$F_s = \pi (4-\pi) + 2\sqrt{\frac{a}{y_n}} \tag{2.4.14}$$

Letting $p = 0.5$ in Eq. (2.4.11), the inverse parabolic section yielded the following equation:

$$F_s = \frac{2y_n}{a} \left\{ \left(1 + 179 \left(\frac{a}{y_n} \right)^3 \right)^{-5} + \left[1 + 0.9 \left(\frac{a}{y_n} \right)^{0.5} \right]^{-5} \right\}^{-0.2} \tag{2.4.15}$$

Drainage layer at a finite depth: Eqs. (2.4.2), (2.4.3), (2.4.4), (2.4.5), (2.4.6), (2.4.7), (2.4.8), (2.4.9), (2.4.10), (2.4.11), (2.4.12), (2.4.13), (2.4.14), and (2.4.15) are applicable for canals passing through a homogeneous and isotropic porous medium of infinite extent. However, in most alluvial plains, the soil is stratified. In many cases, highly permeable layers of sand and gravel underlie the top low permeable layer of finite depth. In that case, the lower layer of sand and gravel acts as a free

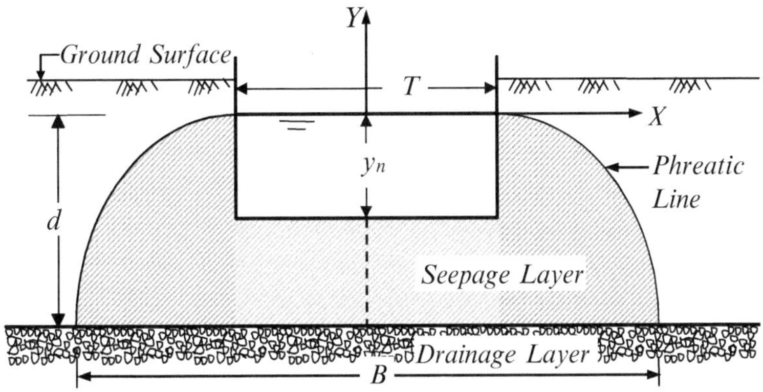

Fig. 2.4 Location of drainage layer

drainage layer for the top seepage layer. The seepage from a canal running through such stratified strata is much more than that in homogeneous medium of very large depth. The difference in quantity of seepage becomes appreciable when the drainage layer lies at a depth less than twice the depth of water in the canal. Swamee et al. (2001a) gave the following simple algebraic equations for the seepage function for polygon canal sections underlain by a drainage layer at depth d (see Fig. 2.4):

$$F_s = \left\{ \left(\frac{1.81m^{1.18} + 2.1}{(d/y_n - 1)^{0.26}} \right)^{9.35} + \left((4\pi - \pi^2)^{1.3} + (2m)^{1.3} \right)^{7.2} \right\}^{0.107}$$

triangular section (2.4.16)

$$F_s = \left\{ \left(\frac{2.5(b/y_n)^{0.84} + 0.45}{(d/y_n - 1)^{0.69}} \right)^{2.38} + \left[(4\pi - \pi^2)^{0.77} + (b/y_n)^{0.77} \right]^{3.094} \right\}^{0.42}$$

rectangular section (2.4.17)

$$F_s = \left\{ \left(1.81 \left[m^{1.3} + 1.432 \left(\frac{b}{y_n} \right)^{0.93} \right]^{0.9} + \frac{b + 100my_n}{2.22b + 47.62my_n + 1.57bm^5} \right)^{p_1} \left(\frac{d}{y_n} - 1 \right)^{-p_1 p_2} + \left\{ \left[(4\pi - \pi^2)^{1.3} + (2m)^{1.3} \right]^{0.77p_3} + \left(\frac{b}{y_n} \right)^{p_3} \right\}^{\frac{p_1}{p_3}} \right\}^{\frac{1}{p_1}}$$

trapezoidal section

(2.4.18)

where

$$p_1 = \frac{2.38b + 7.48my_n}{b + 0.8my_n}; \quad p_2 = \frac{0.318b + 0.26my_n}{0.461b + my_n}; \text{ and } p_3 = \frac{1 + 0.6m}{1.3 + 0.6m} \tag{2.4.19}$$

On the other end, for shallow depth of drainage layer, Swamee and Kashyap (2004) obtained the following equation for a circular channel:

$$F_s = \eta_n^{-1}\left((\eta_n F_{s\infty})^{6.25} + \left\{ \frac{0.87\eta_n^{-0.08} + 1}{[(d/y_n) - 1]^{0.5}} \right\}^{6.25} \right)^{0.16} \tag{2.4.20}$$

where $F_{s\infty}$ = seepage function for circular section with infinite depth of drainage layer given by Eq. (2.4.9). Swamee and Kashyap (2004) gave the following equation for seepage function for a parabolic section:

$$F_s = \frac{y_n^4 B_0 + (d - y_n)^4 B_\infty}{y_n \left[y_n^4 + (d - y_n)^4 \right]} \tag{2.4.21}$$

where

$$B_0 = \frac{1.27a^{1.17} + 2.7y_n^{1.17}}{(d - y_n)^{0.17}} \tag{2.4.22}$$

$$B_\infty = \pi (4 - \pi) y_n + 2\sqrt{ay_n} \tag{2.4.23}$$

As $d \to \infty$, Eqs. (2.4.16), (2.4.17), (2.4.18), (2.4.20), and (2.4.21) become functions of canal geometry only and reduce to the seepage functions for canals passing through a homogeneous porous medium of infinite depth. The drainage layer can be assumed at an infinite depth when $d \geq T + 3y_n$.

2.4.2 Evaporation Loss

Evaporation loss depends on (1) the supply of energy to provide the latent heat of vaporization and (2) the ability to transport the vapor away from the evaporating surface, which in turn depends on the wind velocity over the surface and the specific humidity gradient in the air above the water surface. A large number of equations for estimating evaporative rate are available in the literature. A review indicated that these equations fall into the following categories: (a) energy balance equations, (b) mass transfer equations, and (c) combination of the two. The energy balance equations require a variety of climatological data. The need of sophisticated equipment for direct measurement of radiation, frequent temperature surveys for heat storage, etc., make the method unattractive. On the other hand, the mass transfer equations are most convenient and useful for determining evaporation from flowing canals. The mass transport-type equations are expressed as

$$E = (e_s - e_d)\, f_w \tag{2.4.24}$$

where E = evaporation discharge per unit free surface area, e_s = saturation vapor pressure of the air at the temperature of the water surface, e_d = saturation vapor pressure of the air at the dew point, and f_w = wind function. The difference between the saturation vapor pressure of the air at the temperature of water surface and at the dew point ($e_s - e_d$) was given by Cuenca (1989) as

$$e_s - e_d = 610.78\left[\exp\left(\frac{17.27\theta_w}{237.3 + \theta_w}\right) - R_h\ \exp\left(\frac{17.27\theta_a}{237.3 + \theta_a}\right)\right] \tag{2.4.25}$$

where θ_w = water surface temperature in °C, θ_a = mean air temperature in °C, and R_h = relative humidity expressed as fraction. The wind function for a flowing channel in m/s per Pa was given by Fulford and Sturm (1984) as

$$f_w = 3.704 \times 10^{-11}\,(1 + 0.25u_2) \tag{2.4.26}$$

where u_2 = wind velocity in m/s at 2 m above the water surface. Combining Eqs. (2.4.24), (2.4.25), and (2.4.26), E in m/s is obtained as

$$E = 2.262 \times 10^{-8}\,(1 + 0.25u_2)\left[\exp\left(\frac{17.27\theta_w}{237.3 + \theta_w}\right) - R_h\ \exp\left(\frac{17.27\theta_a}{237.3 + \theta_a}\right)\right] \tag{2.4.27}$$

Equation (2.4.27) shows that in the simplest form of mass transfer approach, E is a function of the wind velocity over the evaporating surface, the water surface temperature, the air temperature, and relative humidity of the air above the water surface, though it may be affected by many other factors. Once E is known, the evaporation loss from a canal can be expressed as

$$q_e = ET \tag{2.4.28}$$

where q_e = evaporation discharge per unit length of canal (m^2/s).

2.4.3 Annual Cost of Total Water Loss

Adding Eqs. (2.4.1) and (2.4.28), the water loss per unit length of canal q_w (m^2/s) becomes

$$q_w = k\,y_n F_s + ET \tag{2.4.29}$$

The water loss m^3/m/year is $60 \times 60 \times 24 \times 365 q_w = 3.1536 \times 10^7 q_w$. Considering c_w as the cost of 1 m^3 of water and using Eq. (2.4.29), the annual cost of water loss (A_w) is $3.1536 \times 10^7 c_w q_w$, thus

$$A_w = 3.1536 \times 10^7 c_w \left(k y_n F_s + ET\right) \tag{2.4.30}$$

It can be seen that whereas C_L and C_e are capital costs, A_w is an annual cost.

2.5 Unification of Costs

The costs of earthwork and lining are incurred at the time of construction of a canal project, whereas the costs of the repair and maintenance of the canal have to be incurred every year. A canal lining may last about 50–60 years. After the life of a component, for example, canal lining, is over, it has to be replaced. The replacement cost has also to be considered as an additional recurring cost. Thus, there are two types of costs: (1) capital cost or the initial investment which has to be incurred for commissioning of the project and (2) the recurring cost which has to be incurred continuously for keeping the project in the operating condition.

These two types of costs cannot be simply added to find the overall cost. These costs have to be brought to same units before they can be added. For combining these costs, the methods generally used are the capitalization method and the annuity method. These methods are described below.

2.5.1 *Capitalization Method*

In this method, the recurring costs are converted to capital costs. This method finds out the amount of money to be kept in a bank yielding an annual interest equal to the annual recurring cost. If an amount C_A is kept in a bank with an annual interest rate of lending r per unit of money, the annual interest on the amount will be rC_A. Equating the annual interest to the annual recurring cost A_r, the capitalized cost C_A is obtained as

$$C_A = \frac{A_r}{r} \tag{2.5.1}$$

A component of a canal network has a finite life T_L. The replacement cost C_R has to be kept in a bank for T_L year so that its interest is sufficient to get the new component. If the original cost of a component is C_0, by selling the component after T_L year as a scrap, an amount αC_0 is recovered, where $\alpha =$ salvage factor. Thus, the net liability after T_L year is

$$C_N = (1 - \alpha)\, C_0 \tag{2.5.2}$$

On the other hand, the amount C_R with interest rate r yields the compound interest I_R given by

$$I_R = \left\{(1 + r)^{T_L} - 1\right\}\, C_R \tag{2.5.3}$$

Equating I_R and C_N, the replacement cost is obtained as

$$C_R = \frac{(1 - \alpha)\, C_0}{(1 + r)^{T_L} - 1} \tag{2.5.4}$$

Denoting the annual maintenance factor as β, the annual maintenance cost is given by βC_0. Using Eq. (2.5.1), the capitalized cost of maintenance C_{ma} works out to be

$$C_{ma} = \frac{\beta C_0}{r} \tag{2.5.5}$$

Adding C_0, C_R, and C_{ma}, the overall capitalized cost C_c is obtained as

$$C_c = C_0\left[1 + \frac{1 - \alpha}{(1 + r)^{T_L} - 1} + \frac{\beta}{r}\right] \tag{2.5.6}$$

Using Eqs. (2.5.1) and (2.5.6), all types of costs can be capitalized to get the overall cost of the project.

Example 2.1 Find the overall cost of canal lining considering scrap factor $\alpha = 0.05$, maintenance factor $\beta = 0.01$, interest rate $r = 0.05$ ₹/₹/year, and life of the lining $T_L = 60$ years.

Solution Using Eq. (2.5.6), the overall cost of lining is

$$C_L = C_{L0}\left[1 + \frac{1 - 0.05}{(1 + 0.05)^{60} - 1} + \frac{0.01}{0.05}\right] = C_{L0}\,(1{+}0.0537{+}0.2) = 1.254 C_{L0}$$

That is, the overall cost of lining is 1.254 times the initial cost.

2.5.2 *Annuity Method*

This method converts the capital costs into recurring costs. The capital investment is assumed to be incurred by borrowing the money that has to be repaid in equal annual installments throughout the life of the component. These installments are paid along with the other recurring costs. The annual installments (called annuity) can be combined with the recurring costs to find the overall annual investment.

If annual installments A_r for the system replacement are deposited in a bank up to T_L year, the first installment grows to $A_r(1+r)^{T_L-1}$, the second installment to $A_r(1+r)^{T_L-2}$, and so on. Thus, all the installments after T_L year add to C_N given by

$$C_N = A_r\left[1 + (1+r) + (1+r)^2 + \cdots + (1+r)^{T_L-1}\right] \tag{2.5.7}$$

Summing up the geometric series, one gets

$$C_N = A_r \frac{(1+r)^{T_L} - 1}{r} \tag{2.5.8}$$

Using Eqs. (2.5.2) and (2.5.8), A_r is obtained as

$$A_r = \frac{(1-\alpha)\,r}{(1+r)^{T_L} - 1} C_0 \tag{2.5.9}$$

The annuity A_0 for the initial investment is given by

$$A_0 = rC_0 \tag{2.5.10}$$

Adding up A_0, A_r, and the annual maintenance cost βC_0, the annuity A is

$$A = rC_0\left[1 + \frac{1-\alpha}{(1+r)^{T_L} - 1} + \frac{\beta}{r}\right] \tag{2.5.11}$$

Comparing Eqs. (2.5.6) and (2.5.10), it can be seen that the annuity is r times the capitalized cost. Thus, one can use either annuity or the capitalization method.

2.5.3 Cost Function

Using Eq. (2.5.1), the annual cost of water loss is converted to the capitalized cost of water loss C_w (₹/m) as

$$C_w = \frac{3.1536 \times 10^7 c_w}{r} (ky_n F_s + ET) \tag{2.5.12}$$

Adding Eqs. (2.2.1), (2.3.1), and (2.5.12), the cost of canal per unit length C (₹/m) is obtained as

$$C = C_e + C_L + C_w = c_e A + c_r A\overline{y} + c_L P + 3.1536 \times 10^7 c_w (ky_n F_s + ET)/r \tag{2.5.13}$$

The following terms are further defined:

$$c_{ws} = 3.1536 \times 10^7 k c_w / r \tag{2.5.14}$$

$$c_{wE} = 3.1536 \times 10^7 E c_w / r \tag{2.5.15}$$

Therefore, Eq. (2.5.13) becomes

$$C = c_e A + c_r A\overline{y} + c_L P + c_{ws} F_s y_n + c_{wE} T \tag{2.5.16}$$

and Eq. (2.2.12) converts to

$$C_w = c_{ws} y_n F_s + c_{wE} T \tag{2.5.17}$$

As c_L/c_e, c_e/c_r, and c_w/c_e have length dimension, they remain unaffected by the monetary unit chosen. These ratios can be obtained for various types of linings, soil strata, and climatic condition by using appropriate unit rates. Using "Schedule" (1997) and "UP" (1992), the c_L/c_e and c_e/c_r ratios were obtained for various types of linings and soil strata. The ratios are listed in Table 2.2.

Table 2.2 Lining and earthwork cost coefficients

Types of strata	c_L/c_e (m)									c_e/c_r (m)
	Type of lining									
	Concrete tile			Brick tile			Brunt clay tile			
	With LDPE film		Without film	With LDPE film		Without film	With LDPE film		Without film	
	100 μ	200 μ		100 μ	200 μ		100 μ	200 μ		
(1)	(2)	(3)	(4)	(5)	(6)	(7)	(8)	(9)	(10)	(11)
Ordinary soil	12.75	13.02	12.24	6.39	6.67	5.88	6.08	6.35	5.57	6.96
Hard soil	10.00	10.22	9.60	5.01	5.23	4.62	4.77	4.99	3.37	8.86
Impure lime nodules	8.90	9.10	8.55	4.47	4.66	4.11	4.25	4.44	3.89	9.96
Dry shoal with shingle	6.56	6.71	6.30	3.29	3.43	3.03	3.13	3.27	2.86	13.50
Slush and lahel	6.40	6.54	6.14	3.21	3.35	2.95	3.05	3.19	2.79	13.86

LDPE low density polyethylene

2.6 Stable Channel Objective Function

A stable channel is a stream in equilibrium that is neither silting nor scouring over a period of time. Obviously, such a stream has developed a cross-sectional area of flow through natural processes of deposition and scour. Lacey (1930) gave a set of empirical equations for flow area, flow perimeter, and bed slope. There has to be an objective function whose minimization yields Lacey's equations. Using Lacey's equations with geometric programming, Swamee (2000) synthesized the following objective function F_E for stable alluvial channels, which is energy spent per unit length of canal:

$$F_E = \frac{\rho g \nu^{1/9} d^{1/3} Q^{8/3}}{3[(s-1)g]^{8/9} A^{5/3} R^{4/3}} + 5.64 \times 10^{-4} \rho g \left\{ \frac{[(s-1)g]^2}{\nu} \right\}^{\frac{1}{3}} Ad + 5.27 \times 10^{-4} \rho g \left\{ \frac{[(s-1)g]^{2.3}}{\nu^{1.6}} \right\}^{\frac{1}{3}} d^{1.1} A^{0.6} R \qquad (2.6.1)$$

where Q = discharge in channel, ρ = mass density of water, g = gravitational acceleration, s = specific gravity of bed material, $R = A/P$ = hydraulic radius, and d = bed material size expressed in m and ν = kinematic viscosity of fluid. The kinematic viscosity depends on the temperature of the fluid, which can be obtained using the equation given by Swamee (2004):

$$\nu = 1.792 \times 10^{-6} \left[1 + \left(\frac{T}{25} \right)^{1.165} \right]^{-1} \qquad (2.6.2)$$

where T = the water temperature in degrees Celsius. The minimum energy per unit length F_E^* corresponding to Eq. (2.6.1) is (Swamee 2000)

$$F_E^* = 0.009023 \rho g \left\{ \frac{[(s-1)g]^4}{\nu^5} \right\}^{\frac{1}{18}} d^{5/6} Q^{5/6} \qquad (2.6.3)$$

For non-alluvial streams, the first term of Eq. (2.6.1), responsible for the flow maintenance, will depend upon the corresponding resistance law. Thus, the general form of the stable channel objective function is expressed as (Swamee 2000)

$$F_E = \rho g Q S_o + 5.64 \times 10^{-4} \rho g \left\{ \frac{[(s-1)g]^2}{\nu} \right\}^{\frac{1}{3}} Ad + 5.27 \times 10^{-4} \rho g \left\{ \frac{[(s-1)g]^{2.3}}{\nu^{1.6}} \right\}^{\frac{1}{3}} d^{1.1} A^{0.6} R \qquad (2.6.4)$$

where S_o = stream bed slope. For river Brahmaputra, F_E^* is found to be (Swamee et al. 2008)

$$F_E^* = 0.01011\rho g\left\{\frac{[(s-1)g]^4}{\nu^5}\right\}^{\frac{1}{18}} d^{5/6}Q^{5/6} \tag{2.6.5}$$

References

Chahar BR (2000) Optimal design of channel sections considering seepage and evaporation losses. Ph.D, thesis, Department of Civil Engineering, University of Roorkee, Roorkee, Uttarakhand, India

Cuenca RH (1989) Irrigation system design: an engineering approach. Prentice Hall, Englewood Cliffs

Fulford JM, Sturm TW (1984) Evaporation from flowing channels. J Energy Eng ASCE 110(1):1–9

Harr ME (1962) Groundwater and seepage. McGraw Hill Book Co. Inc., New York

Indian Bureau of Standard (1980) Measurement of seepage losses from canals. IS-9452, Parts 1 and 2, New Delhi

Lacey G (1930) Sable channel in alluvium. Proc Inst Civil Eng Lond 229:259–285

Morel-Seytoux HJ (1964) Domain variations in channel seepage flow. J Hydraul Eng ASCE 90(2):55–79

Polubarinova-Kochina PY (1962) Theory of ground water movement. Princeton University Press, Princeton

Schedule of Rates (1997) Irrigation Department, Uttar Pradesh, Lucknow, India

Swamee PK (2000) Stable channel objective function. Int J Sediment Res 15(4):434–439

Swamee PK (2004) Improving design guidelines for class-I circular sedimentation tanks. Urban Water J 1(4):309–314. Taylor and Francis

Swamee PK, Kashyap D (2001) Design of minimum seepage-loss nonpolygonal canal sections. J Irrig Drain Eng ASCE 127(2):113–117

Swamee PK, Kashyap D (2004) Design of minimum seepage loss nonpolygon canal sections with drainage layer at shallow depth. J Irrig Drain Eng ASCE 130(2):166–170

Swamee PK, Mishra GC, Chahar BR (2000) Design of minimum seepage loss canal sections. J Irrig Drain Eng ASCE 126(1):28–32

Swamee PK, Mishra GC, Chahar BR (2001a) Design of minimum seepage loss canal sections with drainage layer at shallow depth. J Irrig Drain Eng ASCE 127(5):287–294

Swamee PK, Mishra GC, Chahar BR (2001b) Closure to discussions on 'design of minimum seepage loss canal sections' by Anvar R. Kacimov. J Irrig Drain Eng ASCE 127(3):189–192

Swamee PK, Sharma N, Dwivedi A (2008) Lacey regime equations for river Brahmaputra. J Hydraul Res IAHR 46(5):707–710

U.P. Composite irrigation project: modernization of upper Ganga canal (1992) Investigation and Planning Circle, Aligarh, Uttar Pradesh, India

Wachyan E, Rushton KR (1987) Water losses from irrigation canals. J Hydrol 92(3–4):275–288

Chapter 3
Basic Canal Hydraulics

Abstract Canals are designed for uniform flow considering economy and reliability. Uniform flow is described by a resistance equation. This chapter describes uniform flow equations for viscous flow, turbulent flow, and sediment-transporting channels. Open-channel sections are used for transferring viscous fluids in chemical plants. The Navier-Stokes equations are the governing equations for viscous flow. For steady viscous uniform flow, the Navier-Stokes equation is reduced to two-dimensional form of Poisson's equation, and solution for a rectangular channel has been included in the chapter. For turbulent flow in channels, different uniform flow equations are described with pointing out that Manning's equation is applicable only to the fully rough turbulent flow and in a limited bandwidth of relative roughness. For other flow conditions, a more general resistance equation based on the Colebrook equation is more appropriate. Direct analytic solution of the normal depth in natural/stable channel section is not possible, as the governing equation is implicit and it requires a tedious method of trial and error. Explicit expressions for normal depth associated with viscous flow in rectangular channel and turbulent flow in triangular, rectangular, trapezoidal, circular, and natural channel sections are presented in the chapter. Furthermore, Chap. 3 describes canal operations through normal sluice gate, side sluice gate, and side weir. Moreover, canal discharge measurements using sharp- and broad-crested weirs and linear weir are addressed in the chapter. Finally, the chapter includes explicit critical depth relations for power law and trapezoidal and circular canal sections.

Keywords Uniform flow • Resistance equation • Viscous flow • Turbulent flow • Roughness • Sediment transport • Normal depth • Natural channel • Canal operations • Sluice gate • Side sluice gate • Side weir • Canal discharge measurement • Sharp-crested weir • Broad-crested weir • Linear weir • Critical depth • Power law channel

3.1 Resistance Equations

An open channel is designed to convey the required discharge under uniform flow conditions. A uniform flow is analyzed by using the continuity equation and the resistance equation. The continuity equation for steady flow is

P.K. Swamee, B.R. Chahar, *Design of Canals*, Springer Transactions in Civil and Environmental Engineering, DOI 10.1007/978-81-322-2322-1_3

$$Q = AV \tag{3.1.1}$$

where Q = discharge, A = flow area, and V = average velocity of flow. In this chapter, the resistance equations for viscous flow, turbulent flow, and flow carrying sediments are described.

3.1.1 Viscous Flow in Channels

Open-channel sections are generally adopted in chemical plants for transferring viscous fluids. The type of flow in such cases is generally uniform flow. The Navier-Stokes (N-S) equations are the governing equations for viscous flow. For a channel flow, the N-S equation for incompressible Newtonian fluid in x-direction is

$$\frac{\partial v_x}{\partial t} + v_x \frac{\partial v_x}{\partial x} + v_y \frac{\partial v_x}{\partial y} + v_z \frac{\partial v_x}{\partial z} = X - \frac{1}{\rho} \frac{\partial p}{\partial x} + \nu \nabla^2 v_x \tag{3.1.2}$$

where x, y, and z = coordinate directions; v_x, v_y, and v_z = velocity components in the coordinate directions; X = body force per unit mass in x-direction; and p = pressure. If the x-coordinate direction is assumed along the channel bed in the flow direction, then $X = gS_o$, wherein S_o = longitudinal channel bed slope, which is equal to energy slope in uniform flow. Considering steady-state uniform motion in x-direction and neglecting inertial terms and variation in x-direction, Eq. (3.1.2) reduces to

$$0 = gS_o + \nu \nabla^2 v_x \tag{3.1.3}$$

Rewriting Eq. (3.1.3),

$$\frac{\partial^2 v}{\partial y^2} + \frac{\partial^2 v}{\partial z^2} = -\frac{gS_o}{\nu} \tag{3.1.4}$$

where $v = v_x$ is time mean velocity in x-direction at the point (y, z). Thus, for steady viscous uniform flow, the Navier-Stokes equation is reduced to two-dimensional form of Poisson's equation as given by Eq. (3.1.4). The solution of Eq. (3.1.4) for discharge Q in an open rectangular channel of bed width b as derived by Boussinesq in 1868 and later modified by Cornish (1928) and Woo and Brater (1961) is

$$Q = \frac{gS_o b y_n^3}{3\nu}\left[1 - \frac{384}{\pi^5}\frac{y_n}{b}\sum_{i=0}^{\infty}(2i+1)^{-5}\tanh\left(\frac{2\,i+1}{4}\frac{\pi b}{y_n}\right)\right] \tag{3.1.5}$$

Equation (3.1.5) was experimentally verified by Cornish (1928), Davis and White (1928), Straub et al. (1958), Woo and Brater (1961), and Woener et al. (1968).

3.1.2 Turbulent Flow in Channels

In general, the type of flow in open channels used for various purposes is turbulent. The first resistance equation for turbulent flow region was given by Chézy (Rouse and Ince 1963) in 1769. The Chézy equation (Chow 1973), in its present form, is

$$V = C_z\sqrt{RS_o} \tag{3.1.6}$$

where C_z = Chézy's roughness factor and R = hydraulic radius defined as the ratio of flow area to flow perimeter. The Chézy's factor can be determined by empirical formulae. Commonly used empirical formulae are due to Ganguillet and Kutter and Manning.

In 1869, Ganguillet and Kutter gave a generalized equation based on a variety of experimental data (Rouse and Ince 1963). The Ganguillet and Kutter equation is

$$C_z = \frac{23 + 1/n + 0.00155/S_o}{1 + (23 + 0.00155/S_o)\ n/\sqrt{R}} \tag{3.1.7}$$

where n is a measure of surface roughness known as Kutter's n or presently as Manning's n.

The popularly known Manning's equation is attributed to the Irish engineer Robert Manning. Gauckler (Williams 1970; Rouse 1956) was the first investigator who in 1867 proposed it in the form $V \propto R^{2/3} S_o{}^{1/2}$. In 1876, Hagen obtained the same relationship using Upper Ganga Canal data, which were collected by Cunningham with observations at Roorkee (Cunningham 1880, 1882; Mital 1986).

The Manning's formula, as presently known though misnomer, is

$$V = \frac{1}{n} R^{2/3} S_o^{1/2} \tag{3.1.8}$$

Comparing Eqs. (3.1.6) and (3.1.8),

$$C_z = \frac{1}{n} R^{1/6} \tag{3.1.9}$$

Chow (1973) gave a comprehensive table of values of n for a wide range of conditions. Barnes (1967) compiled a color-photo album to illustrate the natural channel conditions associated with different n values.

"Friction" (1963) recommended the following form of the Darcy-Weisbach equation for uniform flow in open channels:

$$V = \sqrt{\frac{8g}{f} RS_o} \tag{3.1.10}$$

where f = friction factor. The friction factor depends on the Reynolds number and channel surface roughness. Experimental measurements of friction in open channels over a wide range of conditions can be better correlated and understood by using the friction factor in the uniform flow equation. Also, f is commonly used in other areas of engineering, and the experience gained there can be utilized in the open-channel flow.

For high roughnesses and for high enough Reynolds numbers, f is independent of Reynolds number and depends only on the hydraulic radius of the flow for a given type of channel surface. For such a flow condition known as fully rough flow, f is inversely proportional to $R^{1/3}$, resulting in a nearly constant value of n. Hence, Manning's equation is applicable only to the fully rough turbulent flows that, according to Hager (1989), establish when the average roughness height of the channel surface ε satisfies the following condition:

$$\varepsilon \geq 30\nu\left[Q(gS_o)^2\right]^{-0.2} \tag{3.1.11}$$

Furthermore, for applicability of Eq. (3.1.8), the average roughness height should also satisfy (see Christensen 1984)

$$25 \leq \frac{R}{\varepsilon} \leq 250 \tag{3.1.12}$$

For general flow conditions, the Colebrook equation is more appropriate than the Manning's equation. The Colebrook equation for an open-channel flow can be written as ("Friction" 1963)

$$\frac{1}{\sqrt{f}} = -2\log\left(\frac{\varepsilon}{12R} + \frac{0.625\nu}{VR\sqrt{f}}\right) \tag{3.1.13}$$

Table 3.1 lists ε for different channel surfaces ("Friction" 1963).

Swamee (1994) eliminated f between Eqs. (3.1.10) and (3.1.13) to get the following equation:

Table 3.1 Average roughness height for different channel surfaces

Type of channel surface	ε (mm)
Very smooth concrete surface	0.15–0.30
Smooth-troweled concrete surfaces	0.5–0.6
Smooth-finished gunite	0.5–1.5
Ordinary concrete, glazed brickwork	1.16–1.5
Rough concrete	3.0–4.5
Earth channels (straight, uniform)	3.0
Rubble masonry	6.0
Untreated gunite	3.0–10.0

$$V = -2.457\sqrt{gRS_o}\ln\left(\frac{\varepsilon}{12R} + \frac{0.221\nu}{R\sqrt{gRS_o}}\right) \tag{3.1.14}$$

Equation (3.1.14) assumes the same roughness on bed and banks and that velocity isovels follow the contour of channel boundary. Combining Eqs. (3.1.1) and (3.1.14), the following general flow resistance equation is obtained:

$$Q = -2.457A\sqrt{gRS_o}\ln\left(\frac{\varepsilon}{12R} + \frac{0.221\nu}{R\sqrt{gRS_o}}\right) \tag{3.1.15}$$

3.1.3 Sediment-Transporting Canals

At present, there is no reliable resistance equation available for flow in sediment-carrying rigid boundary channels. In the absence of such an equation, the sediment-carrying canals can be designed by using the following equation given by Durand (Stepanoff 1969) for head loss h_f for sediment transport through pipes in heterogeneous suspension:

$$h_f = \frac{fLV^2}{2gD} + \frac{81(s-1)C_v fL\sqrt{(s-1)gD}}{2C_D^{0.75}V} \tag{3.1.16}$$

where L = pipe length, D = pipe diameter, C_v = sediment concentration, and C_D = drag coefficient of particles. Considering $D = 4R$, Eq. (3.1.16) changes to

$$h_f = \frac{fLV^2}{8gR} + \frac{81(s-1)C_v fL\sqrt{(s-1)gR}}{C_D^{0.75}V} \tag{3.1.17}$$

Denoting Q_s as the sediment discharge and assuming Q_s to be small in comparison to Q, the average flow velocity is

$$V = \frac{Q + Q_s}{A} \tag{3.1.18}$$

Similarly, C_v is defined as

$$C_v = \frac{Q_s}{Q + Q_s} \tag{3.1.19}$$

Using Eqs. (3.1.18) and (3.1.19), Eq. (3.1.17) is written as (Swamee and Swamee 2004)

$$h_f = \frac{fL(Q+Q_s)^2}{8A^2 gR} + \frac{81\,(s-1)\,Q_s A f L \sqrt{(s-1)\,gR}}{C_D^{0.75}(Q+Q_s)^2} \tag{3.1.20}$$

Durand (Vanoni 1975) found that the lower limit of the transition, between heterogeneous flow and with moving bed, corresponds fairly accurately to minimum head loss. This velocity has been named as *limit deposit velocity*. For minimum head loss, equating the differential coefficient of Eq. (3.1.20) with respect to Q to zero and simplifying gives

$$Q = 3 \times 2^{0.75} Q_s^{0.25} \left(\frac{A\sqrt{(s-1)\,gR}}{C_D^{0.25}} \right)^{0.75} - Q_s \tag{3.1.21}$$

Substituting Eq. (3.1.21) in Eq. (3.1.20) and simplifying yields

$$Q_s = \frac{2C_D^{0.75} S_o^2}{81(s-1)^{1.5} f^2} A\sqrt{gR} \tag{3.1.22}$$

where $h_f/L = S_o =$ the channel bed slope. Using Eqs. (3.1.21) and (3.1.22),

$$Q = 81 Q_s \left[\frac{(s-1)\,f}{S_o \sqrt{C_D}} \right]^{1.5} - Q_s \tag{3.1.23}$$

The average forward velocity of sediment V_s is given by

$$V_s = \frac{Q_s}{A} \tag{3.1.24}$$

Thus, using Eqs. (3.1.23) and (3.1.24),

$$V_s = \frac{2C_D^{0.75} S_o^2}{81(s-1)^{1.5} f^2} \sqrt{gR} \tag{3.1.25}$$

For f, rewriting Eq. (3.1.13) as

$$f = 1.325 \left[\ln\left(\frac{\varepsilon}{12R} + \frac{2.5}{\mathbf{R}\sqrt{f}} \right) \right]^{-2} \tag{3.1.26}$$

where

$$\mathbf{R} = \frac{4VR}{\nu} \tag{3.1.27}$$

Equation (3.1.26) can be converted to the following explicit form in f (Swamee and Swamee 2004):

$$f = 1.325\left[\ln\left(\frac{\varepsilon}{12R} + \frac{5.69}{\mathbf{R}^{0.9}}\right)\right]^{-2} \tag{3.1.28}$$

For the ranges $10^{-6} \leq \varepsilon/R \leq 10^{-2}$ and $4 \times 10^3 \leq \mathbf{R} \leq 10^8$, the errors involved in the use of Eq. (3.1.28) are within $\pm$ 1 %. Using Eqs. (3.1.25) and (3.1.28),

$$V_s = \frac{C_D^{0.75} S_o^2}{71.1(s-1)^{1.5}} \sqrt{gR}\left[\ln\left(\frac{\varepsilon}{12R} + \frac{5.69}{\mathbf{R}^{0.9}}\right)\right]^4 \tag{3.1.29}$$

3.2 Normal Depth

The depth of flow corresponding to uniform flow is known as the *normal depth*. The normal depth is determined by the resistance equation. Normal depth is a key parameter occurring in the design of canals. For practical canal sections, the determination of normal depth involves solution of implicit equations. Direct analytic solution of the normal depth in natural/stable channel section is not possible, as the governing equation is implicit. The solution of the implicit equations requires tedious method of trial and error.

3.2.1 *Viscous Flow in Rectangular Channel*

Equation (3.1.5) is written as

$$M_b = \frac{\beta_n^3}{3}\left[1 - \frac{384}{\pi^5}\beta_n \sum_{i=0}^{\infty} (2i+1)^{-5} \tanh\left(\frac{\pi}{4}\frac{2i+1}{\beta_n}\right)\right] \tag{3.2.1}$$

where $M_b = \nu Q/\left(gS_o b^4\right)$ and $\beta_n = y_n/b$. Swamee and Chahar (2009) have obtained the following explicit equation for Eq. (3.2.1):

$$y_n = 1.55bM_b^{0.337}\frac{1 + 10.01M_b^{0.881}}{1 + 0.391M_b^{0.645}} \tag{3.2.2}$$

Equation (3.2.2) is sufficiently accurate for all practical purposes. For given parameters (discharge, bed width, bed slope, and viscosity of fluid), the dimensionless parameter M_b can be computed, and then Eq. (3.2.2) results the normal depth in single-step computation involving a negligible error.

It should be noted that Eq. (3.1.5) and hence Eqs. (3.2.1) and (3.2.2) are applicable in the laminar flow range only. The upper limit of Reynolds number for laminar flow in open channel is 2,000; therefore, the range of validity of these equations is

$$\frac{4Vby_n}{\nu\,(b+2y_n)} < 2,000 \tag{3.2.3}$$

Example 3.1 Determine the normal depth in 0.5 m wide rectangular channel section for carrying a discharge of 0.025 m³/s on a bed slope of 0.005. The kinematic viscosity of fluid is 4×10^{-3} m²/s.

Adopting $g = 9.79$ m/s² and then using definition of M_b, $M_b = 0.0327$. Solving Eq. (3.1.5) by trial-and-error method, $\beta_n = 0.6997$; hence, $y_n = 0.3499$ m, while Eq. (3.2.2) directly yields $y_n = 0.3499$ m. Thus, there is no error.

3.2.2 *Turbulent Flow Channels*

3.2.2.1 Triangular Section

The resistance Eq. (3.1.15) for a triangular section reduces to

$$Q = -1.737\frac{m\sqrt{mgS_o}}{(1+m^2)^{0.25}}y_n^{2.5}\ln\left[\frac{\varepsilon\sqrt{1+m^2}}{6my_n} + \frac{0.625\nu(1+m^2)^{0.75}}{my_n\sqrt{gmy_nS_o}}\right] \tag{3.2.4}$$

For a hydrodynamically rough channel, the following exact solution is obtained (Swamee and Rathie 2012):

$$y_n = 1.157\frac{(1+m^2)^{0.1}Q^{0.4}}{m^{0.6}(gS_o)^{0.2}}\left\{W_0\left[\frac{127mQ}{(1+m^2)\,\varepsilon^2\sqrt{g\varepsilon S_o}}\right]\right\}^{-0.4} \tag{3.2.5}$$

where W_0 is the principal branch of Lambert's W function. See Appendix 1 for an introduction of Lambert's W function. On the other hand, for a hydrodynamically smooth channel, the exact solution is (Swamee and Rathie 2012)

$$y_n = 0.9836\frac{(1+m^2)^{0.1}Q^{0.4}}{m^{0.6}(gS_o)^{0.2}}\left\{W_0\left[2.1mQ\left(\frac{gS_o}{\nu^5\sqrt{1+m^2}}\right)^{\frac{1}{3}}\right]\right\}^{-0.4} \tag{3.2.6}$$

Swamee (1994) found the following general equation for normal depth, which is valid for both rough and smooth channels:

$$y_n = 0.468\left[\frac{\varepsilon(1+m^2)^{2.9}}{m^{15.4}}\left(\frac{Q}{\sqrt{gS_o}}\right)^{9.6} + \frac{7\nu(1+m^2)^{3.1}}{m^{15.6}}\frac{Q^{9.4}}{(gS_o)^{5.2}}\right]^{0.04} \tag{3.2.7}$$

The maximum error involved in Eq. (3.2.7) is 1.5 %, which fully justifies its use for computation of normal depth.

3.2.2.2 Rectangular Section

For a rectangular section, Eq. (3.1.15) becomes

$$Q = -2.457\sqrt{\frac{b^3 y_n^3 g S_o}{b + 2y_n}} \ln\left[\frac{\varepsilon\,(b + 2y_n)}{12by_n} + \frac{0.221\nu}{\sqrt{gS_o}}\left(\frac{b + 2y_n}{by_n}\right)^{1.5}\right] \tag{3.2.8}$$

For a narrow rectangular channel (b is very small in comparison to y_n, i.e., $b/y_n \to 0$), Eq. (3.2.8) leads to the following closed-form solution for y_n:

$$y_n = \frac{0.576Q}{b\sqrt{gbS_o}}\left[\ln\left(\frac{6b\sqrt{gbS_o}}{\varepsilon\sqrt{gbS_o} + 3.75\nu}\right)\right]^{-1} \tag{3.2.9}$$

On the other hand, for a wide rectangular channel (b is very large in comparison to y_n, i.e., $b/y_n \to \infty$), Swamee and Rathie (2012) gave the following equations for rough and smooth surfaces, respectively:

$$y_n = 0.72\left[\frac{b\sqrt{gS_o}}{Q} W_0\left(\frac{25.378Q}{b\varepsilon\sqrt{\varepsilon g S_o}}\right)\right]^{-\frac{2}{7}} \tag{3.2.10}$$

$$y_n = 0.5492\left[\frac{b\sqrt{gS_o}}{Q} W_0\left(\frac{1.842Q}{b\nu}\right)\right]^{-\frac{2}{7}} \tag{3.2.11}$$

Furthermore, Swamee (1994) fitted the following equation for a wide rectangular canal:

$$y_n = 0.213\left[\varepsilon\left(\frac{Q}{b\sqrt{gS_0}}\right)^{10.444} + 70\nu^{1.2}\left[\frac{Q}{b(gS_0)^{0.561}}\right]^{9.911}\right]^{0.06} \tag{3.2.12}$$

The maximum error involved in the use of Eq. (3.2.12) is 1 %. Combining Eqs. (3.2.9) and (3.2.12), Swamee (1994) obtained the following explicit equation for normal depth in any rectangular channel:

$$\begin{aligned} y_n = 0.213 &\left(\left\{\varepsilon\left(\frac{Q}{b\sqrt{gS_0}}\right)^{10.444} + 70\nu^{1.2}\left[\frac{Q}{b(gS_0)^{0.561}}\right]^{9.911}\right\}^{0.12}\right. \\ &+ \left.\left\{7.3\frac{Q^2}{gb^3S_o}\left[\ln\left(\frac{6b\sqrt{gbS_o}}{\varepsilon\sqrt{gbS_o}+3.75\nu}\right)\right]^{-2}\right\}\right)^{0.5} \end{aligned} \tag{3.2.13}$$

The maximum error involved in the use of Eq. (3.2.13) is 3 %, which is within practical limits.

3.2.2.3 Trapezoidal Section

For a very small depth of flow, the trapezoidal channel can be regarded as a wide rectangular channel, whereas for a very large flow depth, it behaves as a triangular channel. Therefore, by combining Eqs. (3.2.7) and (3.2.12), Swamee (1994) obtained the following explicit equation for the normal depth in a trapezoidal channel:

$$y_n = 0.213\left(\left\{\varepsilon\left(\frac{Q}{b\sqrt{gS_0}}\right)^{10.444} + 70\nu^{1.2}\left[\frac{Q}{b(gS_0)^{0.561}}\right]^{9.911}\right\}^{-0.06/p} + 0.455^{1/p}\left[\frac{\varepsilon\left(1+m^2\right)^{2.9}}{m^{15.4}}\left(\frac{Q}{\sqrt{gS_o}}\right)^{9.6} + \frac{7\nu\left(1+m^2\right)^{3.1}}{m^{15.6}}\frac{Q^{9.4}}{(gS_o)^{5.2}}\right]^{-0.04/p}\right)^{-p} \quad (3.2.14)$$

where $p = 0.45\,m^{0.25}$ and $0.5 \leq m \leq 3.5$. One can expect maximum error about 5 % in use of Eq. (3.2.14) in the stated range of m.

3.2.2.4 Circular Section

Using Eq. (3.1.15), the following equation for the normal depth was obtained (Swamee and Swamee 2008):

$$y_n = \frac{0.943D}{0.823\left(\frac{Q_{\max}}{Q} - 1\right)^{0.665} + 1} \quad (3.2.15a)$$

where Q_{max} = maximum discharge occurring at $y_n = 0.943D$ which is given by (Swamee and Swamee 2008)

$$Q_{\max} = -D^2\sqrt{gDS_o}\ln\left(\frac{\varepsilon}{3.46D} + \frac{1.43\nu}{D\sqrt{gDS_o}}\right) \quad (3.2.15b)$$

Barring small depths $y_n \leq 0.3D$, the maximum error involved in Eq. (3.2.15a, b) is well within 1 %.

3.2.3 *Natural Channels*

Natural streams are self-formed channels made by delicate balance of natural processes of scour and deposition. Channel form is dictated by independent variables of hydrologic discharge, sediment supply, and character of boundary materials, including vegetation. Dependent variables are those physical characteristics that define channel form (cross-section slope and plan-form). In a stable reach of a natural channel, the flow is uniform and the corresponding depth of flow is normal.

3.2.3.1 Equation of Natural Channel Section

Natural channel design involves determining the dimension, pattern, and profile of a stable reach cross section. Stability is defined as the ability of a stream, over time, to transport the flows and sediments of the watershed while neither aggrading nor degrading and while maintaining a consistent dimension, pattern, and profile. Channels must be shaped as closely as possible to a naturally stable state to minimize post-construction adjustment. Henderson (1966) gave the following equation of a stable channel section:

$$y = y_n \cos\left(\frac{x \tan\varphi}{y_n}\right) \tag{3.2.16}$$

where x and y = horizontal and vertical coordinates, respectively, and φ = angle of repose of bed material. Using Eq. (3.2.16), the following geometric properties were obtained by Henderson (1966):

$$A = \frac{2y_n^2}{\tan\varphi} \tag{3.2.17}$$

$$P = \frac{2y_n E}{\sin\varphi} \tag{3.2.18}$$

$$R = \frac{y_n \cos\varphi}{E} \tag{3.2.19}$$

where E = complete elliptical integral of the second kind (Abramowitz and Stegun 1972) defined as

$$E = \int_0^{\pi/2} \sqrt{1 - \sin^2\varphi \sin^2\theta}\, d\theta \tag{3.2.20}$$

3.2.3.2 Normal Depth Equation

Using Eqs. (3.2.17) and (3.2.19), Eq. (3.1.15) changes to

$$y_{n*} = 0.529\left[-\ln\left(\frac{\varepsilon_*}{12y_{n*}} + \frac{0.221\nu_*}{y_{n*}^{1.5}}\right)\right]^{-0.4} \tag{3.2.21}$$

where

$$y_{n*} = y_n \left(\frac{g S_o \cos^3\varphi}{Q^2 E \sin^2\varphi} \right)^{0.2} \tag{3.2.22}$$

$$\varepsilon_* = \varepsilon \left(\frac{g S_o E^4}{Q^2 \sin^2\varphi \cos^2\varphi} \right)^{0.2} \tag{3.2.23}$$

$$\nu_* = \nu \left(\frac{E^6}{g S_o Q^3 \cos^3\varphi \sin^3\varphi} \right)^{0.2} \tag{3.2.24}$$

The Reynolds number **R** is defined as

$$\mathbf{R} = \frac{4RQ}{\nu A} \tag{3.2.25}$$

Using Eqs. (3.2.17), (3.2.19), and (3.2.22), Eq. (3.2.25) simplifies to

$$\mathbf{R} = \frac{2}{\nu_* y_{n*}} \tag{3.2.26}$$

Swamee and Chahar (2010) fitted the following explicit equation to Eq. (3.2.21) for y_{n*}:

$$y_{n*} = 0.31(\varepsilon_* + 7\nu_*)^{0.04} \tag{3.2.27}$$

Equation (3.2.27) is valid for $10^{-8} \le \nu_* \le 10^{-3}$ and $10^{-6} \le \varepsilon_* \le 10^{-2}$ which correspond to $10^4 \le \mathbf{R} \le 10^8$ and $5 \times 10^{-6} \le \varepsilon/y_n \le 2 \times 10^{-3}$. For these ranges of ν_* and ε_*, Eq. (3.2.27) was checked for accuracy with Eq. (3.2.21), and it was found to yield y_{n*} within an error of $\pm$ 3.25 %. Equation (3.2.27) is converted to the following directly usable form:

$$y_n = 0.31 \left[\frac{\varepsilon E}{\cos\varphi} \left(\frac{Q^2 E \sin^2\varphi}{g S_o \cos^3\varphi} \right)^{4.8} + \frac{7\nu E}{Q \sin\varphi} \left(\frac{Q^2 E \sin^2\varphi}{g S_o \cos^3\varphi} \right)^{5.2} \right]^{0.04} \tag{3.2.28}$$

Putting $\nu = 0$ for rough turbulent flow, Eq. (3.2.28) reduces to

$$y_n = 0.31 \left(\frac{\varepsilon E}{\cos\varphi} \right)^{0.04} \left(\frac{Q^2 E \sin^2\varphi}{g S_o \cos^3\varphi} \right)^{0.192} \tag{3.2.29}$$

On the other hand, putting $\varepsilon = 0$ for smooth turbulent flow, Eq. (3.2.28) changes to

$$y_n = 0.335 \left(\frac{\nu E}{Q \sin\varphi} \right)^{0.04} \left(\frac{Q^2 E \sin^2\varphi}{g S_o \cos^3\varphi} \right)^{0.208} \tag{3.2.30}$$

3.2.3.3 Approximation to Elliptical Integral

Equations (3.2.28), (3.2.29), and (3.2.30) involve complete elliptical integral of the second kind defined by Eq. (3.2.20), which requires numerical integration or double interpolations from tabulated values (Abramowitz and Stegun 1972). Adopting a curve fitting method, the following simple equation has been fitted by Swamee and Chahar (2010):

$$E = \frac{\frac{\pi}{2}\left(\frac{\pi}{2} - \varphi\right)^{1.462} + 0.622\varphi^{1.462}}{\left(\frac{\pi}{2} - \varphi\right)^{1.462} + 0.622\varphi^{1.462}} \tag{3.2.31}$$

where φ is in radians. Equation (3.2.31) is near exact as the involved absolute error is less than 0.34 %. Equation (3.2.31) avoids complexities of the numerical integration or inconvenience in interpolations of the tabulated values, and it is useful in other cases where complete elliptical integral of the second kind appears in solution.

Example 3.2 Determine the dimensions of a stable channel section for carrying a discharge of 200 m^3/s on a bed slope of 0.0001. Assume the kinematic viscosity of water = 10^{-6} m^2/s, the average height of roughness of channel surface = 3 mm, and angle of repose of channel material = $\pi/6$.

$E = 1.4675$ for $\phi = \pi/6$. Adopting $g = 9.79$ m/s^2 and then using Eqs. (3.2.23) and (3.2.24), $\varepsilon_* = 1.7121 \times 10^{-4}$ and $\nu_* = 4.3572 \times 10^{-7}$, respectively. Solving Eq. (3.2.21) by trial and error, $y_{n*} = 0.2141$; hence, from Eq. (3.2.22), $y_n = 6.3556$ m, while Eq. (3.2.28) directly yields $y_{n*} = 0.2193$ or $y_n = 6.5110$ m. Thus, the error in the normal depth computation = 2.445 %, which is negligible. Using this normal depth, $A = 139.9286$ m^2, $P = 37.3065$ m, $R = 3.7508$ m, and average velocity of flow = 1.4293 m/s. Furthermore, using Eq. (3.2.31), $E = 1.4657$, so error = 0.123 %.

3.2.3.4 Limitations

Various methods to find the stable dimensions of natural channels vary widely because of varying geomorphological conditions and sediment transport rates. For example, $P = 68.8$ m and $R = 1.206$ m, if Lacey's regime equations are used for the same data. Similar or more variations in natural channel dimensions may be there with equations by other researchers. The proposed method is based on Henderson's cosine profile for a stable channel and should be used with the following limitations of the Henderson's cosine profile. In the cosine profile of natural channel, every point of the periphery is assumed to be at critical equilibrium condition, i.e., the shear stress at the point is equal to critical tractive stress. Such condition may not exist in a natural channel as there is wide variation in critical state from point to point. In practical solutions, sediment transport rate must be included. In the proposed equation, only skin roughness is considered, whereas, in natural channels, there will be form roughness as well. Also, the equation is not valid for natural channels of meandering and braided type.

3.3 Canal Operations

For various reasons, discharge flowing in a main canal or distributary has to be changed. This task is accomplished by a sluice gate. Discharge in a main canal is varied by a sluice gate placed normal to the flow direction. On the other hand, a part of discharge of a main canal is diverted into a branch canal by a side sluice gate or an oblique sluice gate.

3.3.1 *Normal Sluice Gate*

A sluice gate is an opening in a hydraulic structure used for controlling the discharge. Figure 3.1 shows flow through a sluice gate with no side or bottom contraction. Downstream free flow occurs at a (relatively) large ratio of upstream depth to the gate-opening height. However, submerged flow at the downstream would occur for low values of this ratio. For a freely issuing stream from a sluice gate, the water surface is quite smooth, whereas for a submerged flow, the corresponding flow profile is extremely rough.

The conventional sluice-gate discharge equation is written in the form

$$Q = C_d ab\sqrt{2gh_0} \tag{3.3.1}$$

where Q = the sluice-gate discharge, a = the sluice-gate opening, b = the sluice-gate length, h_0 = the upstream water depth, g = gravitational acceleration, and C_d = discharge coefficient. Figure 3.2 shows the variation of C_d under free and submerged flow conditions as obtained by Henry (1950). Henry's experimental investigation is considered most extensive and reliable. Henry's investigation was later confirmed by Rajaratnam and Subramanya (1967).

3.3.1.1 Free Flow Condition

A perusal of Fig. 3.2 indicates for free flow; the discharge coefficient progressively increases to a saturation value of 0.611. Hydraulically the sluice gate ceases to exist when $h_0 = a$ or less. Thus, for $h_o = a$, the discharge coefficient $C_d = 0$. Considering this limiting value and the coordinates of the free flow discharge curve, the following equation is obtained (Swamee 1992):

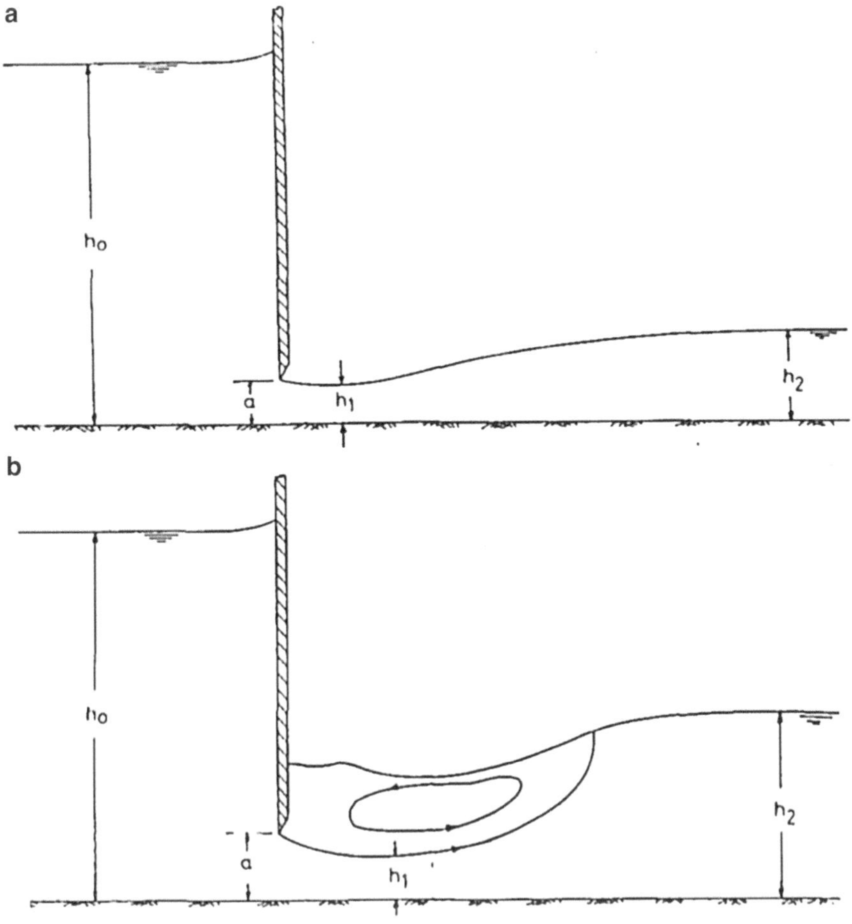

Fig. 3.1 Definition sketch: (**a**) free flow and (**b**) submerged flow

$$C_d = 0.611\left(\frac{h_0 - a}{h_0 + 15a}\right)^{0.072} \tag{3.3.2}$$

Equation (3.3.2) has an excellent agreement with the experimental curve of Fig. 3.2. Using Eqs. (3.3.1) and (3.3.2), the following equation for the sluice-gate discharge is obtained:

$$Q = 0.864ab\sqrt{gh_0}\left(\frac{h_0 - a}{h_0 + 15a}\right)^{0.072} \tag{3.3.3}$$

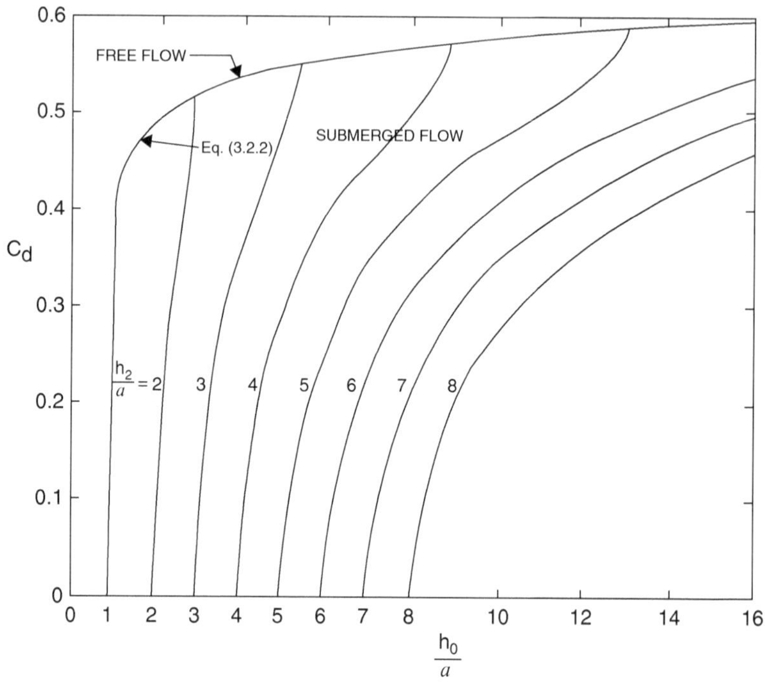

Fig. 3.2 Variation of discharge coefficient

3.3.1.2 Submerged Flow Condition

Under the submerged flow condition, the discharge coefficient is zero when $h_0 = h_1$ (the tail-water depth). Any increase in h_0 above h_2 results in rapid increase in discharge coefficient until h_0 attains a maximum value $h_{0\max}$ at which the flow is free. From Fig. 3.2, $h_{0\max}/a$ can be obtained for various values of h_2/a. Plotting these values on a double logarithmic paper, the following equation is obtained:

$$\frac{h_{0\ \max}}{a} = 0.81\left(\frac{h_2}{a}\right)^{1.72} \tag{3.3.4}$$

So long as h_0 lies between h_2 and $h_{0\max}$, submerged flow conditions will prevail. That is, for submerged flow to exist, the following condition is to be satisfied:

$$h_2 < h_0 < 0.81 h_2\left(\frac{h_2}{a}\right)^{1.72} \tag{3.3.5}$$

On the other hand, the condition for existence of free flow is

$$h_0 \geq 0.81 h_2 \left(\frac{h_2}{a} \right)^{0.72} \tag{3.3.6}$$

Using the discharge coefficient curves of Fig. 3.2 for submerged flow, the following equation has been fitted to describe the curves:

$$C_d = 0.611 \left(\frac{h_0 - a}{h_0 + 15a} \right)^{0.072} (h_0 - h_2)^{0.7} \left\{ 0.32 \left[0.81 h_2 \left(\frac{h_2}{a} \right)^{0.72} - h_0 \right]^{0.7} + (h_0 - h_2)^{0.7} \right\}^{-1} \tag{3.3.7}$$

Equation (3.3.7) is valid for the conditions described in Eq. (3.3.5). It can be seen that for $h_0 = h_{0\max}$, Eq. (3.3.7) reduces to Eq. (3.3.2), which is the upper limit of its applicability. Similarly at the lower limit of its applicability for $h_0 = h_2$, Eq. (3.3.7) yields $C_d = 0$. For intermediate values of h_0, the difference between the values obtained by Eq. (3.3.7) and that from Fig. 3.2 is negligible. Combining Eqs. (3.3.1) and (3.3.7), the sluice-gate discharge equation is given by

$$Q = 0.864 ab \sqrt{g h_0} \left(\frac{h_0 - a}{h_0 + 15a} \right)^{0.072} (h_0 - h_2)^{0.7} \left\{ 0.32 \left[0.81 h_2 \left(\frac{h_2}{a} \right)^{0.72} - h_0 \right]^{0.7} + (h_0 - h_2)^{0.7} \right\}^{-1} \tag{3.3.8}$$

3.3.1.3 Practical Applications

The methodology developed in the preceding sections can be used to solve the following commonly occurring problems in the sluice-gate operation:

1. To determine whether the flow is free or submerged (this problem can be solved by application of Eqs. (3.3.5) and (3.3.6))
2. To predict the discharge (the discharge can be predicted by application of Eq. (3.3.3) for free flow and Eq. (3.3.8) for submerged flow as per flow conditions determined earlier)
3. To decide the gate opening for a given discharge (in this case, flow can be either free or submerged. Take an arbitrary gate opening and use Eqs. (3.3.5) and (3.3.6) to find the flow condition and accordingly calculate the discharge using either Eq. (3.3.3) (for free flow) or Eq. (3.3.8) (for submerged flow). If the calculated discharge is greater than the given discharge, reduce the gate opening; otherwise increase it, and repeat the process until the difference between the calculated and the given discharge is small)

4. To decide the upstream depth for a given discharge (take a trial value of h_0, and find the flow condition using Eqs. (3.3.5) and (3.3.6) to accordingly obtain the discharge using Eq. (3.3.3) or Eq. (3.3.8). If the calculated discharge is more than the given discharge, reduce the upstream depth; otherwise, increase it and repeat the process until the difference between the calculated and given discharge is small)
5. To decide the tail-water depth for a given discharge (the tail-water depth has to be increased if the free flow discharge is larger than the given discharge. Take a trial value of h_2, ensuring the submerged flow condition described by Eq. (3.3.5), and calculate the discharge using Eq. (3.3.8). If the calculated discharge is more than the given discharge, increase the tail-water depth; otherwise, reduce it, such that the condition of Eq. (3.3.5) is followed, and repeat the process until the difference of the calculated and the given discharge is small).

These algorithms can be adopted on computer-controlled sluice-gate operations in an irrigation canal network.

3.3.2 *Side Sluice Gate*

A side sluice gate is a rectangular opening (created by a vertical sliding gate) in the side of a canal through which lateral outflow into a side channel takes place. Side sluice gates are widely used as head regulators for canals, branches, silt flushing in a power canal forebay, and so on. Flow through a side sluice gate is governed by the equation of spatially varied flow with decreasing discharge. See Fig. 3.3.

For the flow through a side sluice gate, the governing differential equation for a rectangular prismatic channel is

$$\frac{dy}{dx} = \frac{S_o - S_f - \frac{Q}{gA^2}\frac{dQ}{dx}}{1 - \frac{Q^2T}{gA^3}} \tag{3.3.9}$$

where S_f = friction slope. Equation (3.3.9) assumes insignificant energy loss from flow diversion. For subcritical flow, Eq. (3.3.9) indicates that there is normally a rising flow profile. Considering the discharge dQ through an elementary strip of length dx along the sluice gate (see Fig. 3.3), one gets

$$\frac{dQ}{dx} = -C_e a\sqrt{2gy} \tag{3.3.10}$$

where C_e = elementary discharge coefficient of the strip. Using Eq. (3.1.15), Swamee (2002) has given the following equation for friction slope:

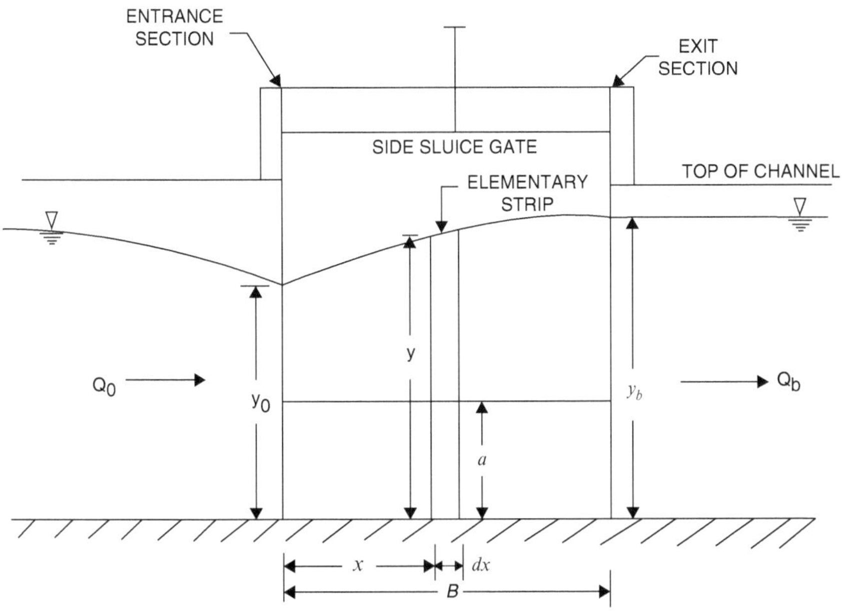

Fig. 3.3 Definition sketch

$$S_f = \frac{Q^2 P}{6.04 g A^3}\left\{-\ln\left[\frac{\varepsilon}{12R} + 1.63\left(\frac{\nu P}{Q}\right)^{0.9}\right]\right\}^{-2} \tag{3.3.11}$$

For a rough rectangular canal, Eq. (3.3.11) simplifies to

$$S_f = \frac{Q^2 (b + 2y)}{6.04 g b^3 y^3}\left\{-\ln\left[\frac{\varepsilon (b + 2y)}{12by}\right]\right\}^{-2} \tag{3.3.12}$$

For a rectangular channel section of bed width b, substitution of Eqs. (3.3.10) and (3.3.12) into Eq. (3.3.9) yields

$$\frac{dy}{dx} = \frac{S_o - \frac{Q^2(b+2y)}{6gb^3y^3}\left\{-\ln\left[\frac{\varepsilon(b+2y)}{12by}\right]\right\}^{-2} + \frac{Q}{gb^3y^2} C_e a\sqrt{2gy}}{1 - \frac{Q^2}{gb^2y^3}} \tag{3.3.13}$$

The elementary discharge coefficient is given by the following (Swamee et al. 1993):

For free flow:

$$C_e = 0.611\left(1 + 0.0112\frac{c}{a}\right)\left(\frac{y - a}{y + a}\right)^{0.216} \tag{3.3.14}$$

For submerged flow:

$$C_e = 0.611\left(1 + 0.0112\tfrac{c}{a}\right)\left(\tfrac{y-a}{y+a}\right)^{0.216}$$
$$.(y - y_t)^{0.67}\left\{\tfrac{0.24}{(1+0.05c/a)}\left[2.5\left(1 + 0.0188\tfrac{c}{a}\right)y_t\left(\tfrac{y_t}{a}\right)^{0.2} - y\right]^{0.67} + (y - y_t)^{0.67}\right\}^{-1} \tag{3.3.15}$$

where c = sluice-gate lip width and y_t = tail-water depth.

Equations (3.3.10) and (3.3.13) can be solved as an initial-value problem with the following prescribed conditions: $x = 0$, $y = y_0$, and $Q = Q_0$, where y_0 = flow depth at the entrance section and Q_0 = entrance-section discharge.

3.3.2.1 Submerged Flow Condition

It can be seen from Eq. (3.3.15) that for $y = y_t$, $C_e = 0$. An increase of y above y_t causes rapid increase in C_e, until y attains a value predicted by Eq. (3.3.14) for free flow condition. Thus, y attains a maximum value y_{max} given by

$$y_{\max} = 2.5\left(1 + 0.0188\frac{c}{a}\right)y_t\left(\frac{y_t}{a}\right)^{0.2} \tag{3.3.16}$$

At which the submerged flow condition has just ended; and the flow is free. Thus, the condition for existence of submerged flow is given by the following:

$$y_t < y < 2.5\left(1 + 0.0188\frac{c}{a}\right)y_t\left(\frac{y_t}{a}\right)^{0.2} \tag{3.3.17}$$

Similarly, the condition for the existence of free flow is as follows:

$$y \geq 2.5\left(1 + 0.0188\frac{c}{a}\right)y_t\left(\frac{y_t}{a}\right)^{0.2} \tag{3.3.18}$$

3.3.3 *Side Weir*

Side weir is an overflow weir provided in a side of a channel that allows lateral flow when the water surface in the main channel rises above the weir crest. In irrigation engineering side weirs with broad crests are used as head regulators of distributaries and escapes. On the other hand, side weirs are widely used for storm overflow from urban drainage systems.

Considering the discharge dQ through an elementary strip of length dx along the side weir (see Fig. 3.4), one gets

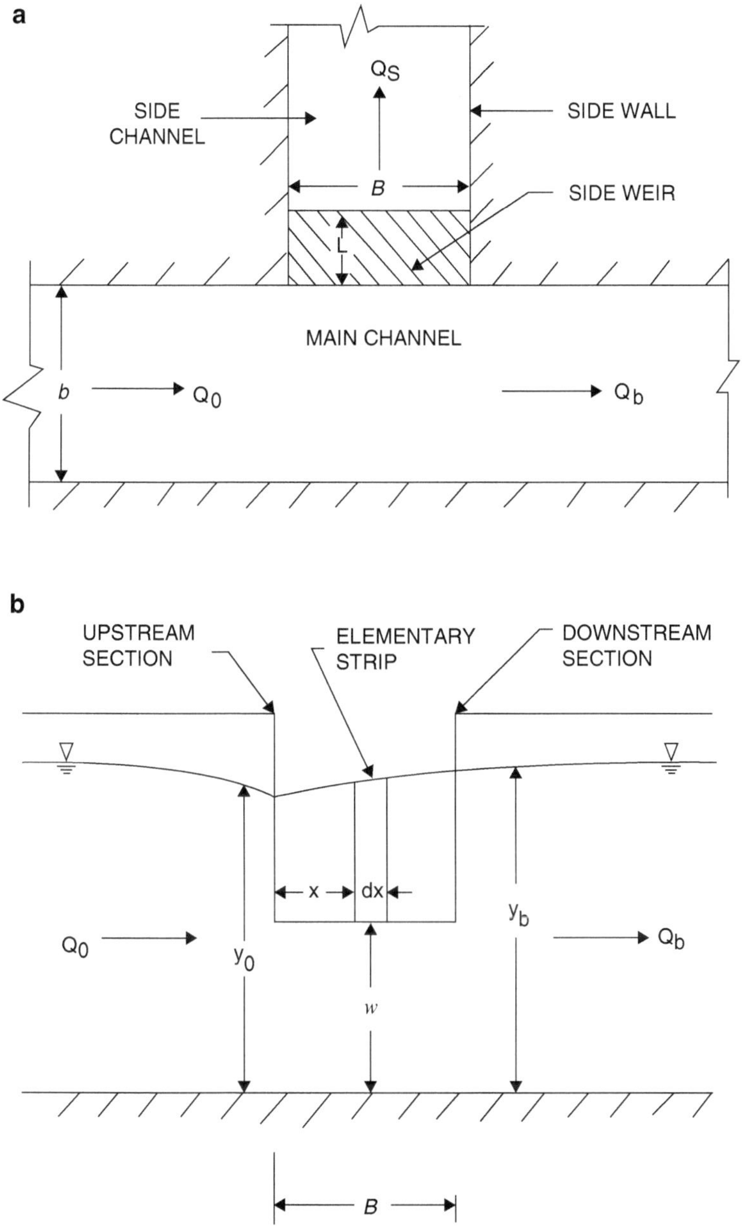

Fig. 3.4 Definition sketch: (**a**) plan and (**b**) elevation

$$\frac{dQ}{dx} = -\frac{2}{3} C_e (y - w) \sqrt{2g(y - w)} \tag{3.3.19}$$

Similar to Eq. (3.3.13), the governing differential equation is given by

$$\frac{dy}{dx} = \frac{S_o - \frac{Q^2(b+2y)}{6gb^3y^3}\left\{-\ln\left[\frac{\varepsilon(b+2y)}{12by}\right]\right\}^{-2} + \frac{2\sqrt{2}Q}{3b^2y^2} C_e (y - w)\sqrt{\frac{y-w}{g}}}{1 - \frac{Q^2}{gb^2y^3}} \tag{3.3.20}$$

where w = weir height (see Fig. 3.4). The elementary discharge coefficient is given by Swamee et al. (1994) for the following cases:

Unrestricted outflow (there are no walls in the side channel and the jet of water is unconstrained)

Sharp-crested weirs:

$$C_d = 0.447\left[\left(\frac{44.7}{50 + \eta_w}\right)^{6.67} + \left(\frac{\eta_w}{1 + \eta_w}\right)^{6.67}\right]^{-0.15} \tag{3.3.21}$$

Broad-crested weirs:

$$C_d = 0.245 + 0.1\frac{\eta_L^{3.3} + 0.025\eta_L^7}{1 + 5.5\eta_L^{0.02}} \tag{3.3.22}$$

where $\eta_w = (y - w)/w$, $\eta_L = (y - w)/L$, and L = weir crest width (see Fig. 3.4).

Restricted outflow (there are side walls in the side channel and the jet of water is constrained by the side walls)

Sharp-crested weirs:

$$C_d = 0.465\left[\left(\frac{46.5}{41.1 + \eta_w}\right)^{10} + \left(\frac{\eta_w}{1 + \eta_w}\right)^{10}\right]^{-0.1} \tag{3.3.23}$$

Broad-crested weirs:

$$C_d = 0.447 + 0.1\frac{\eta_L^{1.79} + 0.05\eta_L^{1.69}}{1 + 2.9\eta_L^{0.02}} \tag{3.3.24}$$

Equations (3.3.19) and (3.3.20) can be solved as an initial-value problem with the following prescribed conditions: $x = 0$, $y = y_0$, and $Q = Q_0$, where y_0 = flow depth at the entrance section and Q_0 = entrance-section discharge.

3.4 Canal Discharge Measurements

3.4.1 Rectangular Weir

3.4.1.1 Sharp-Crested Weir

Weir is a reliable discharge-measuring device, which is nothing but a raised obstruction in the flow. See Fig. 3.5.

Rectangular Weir: The conventional form of the weir discharge equation is

$$Q = \frac{2}{3} C_d b h \sqrt{2gh} \tag{3.4.1}$$

where $b =$ channel width ($=$ weir length), $h =$ head on the weir crest (as shown in Fig. 3.3), and $C_d =$ discharge coefficient. Swamee (1988) gave the following full range equation for the discharge coefficient:

$$C_d = 1.06 \left[\left(\frac{14.14w}{8.15w + h} \right)^{10} + \left(\frac{h}{w + h} \right)^{15} \right]^{-0.1} \tag{3.4.2}$$

where $w =$ the weir height. See Fig. 3.5.

3.4.1.2 Broad-Crested Weir

The weir equation can be further generalized to include the effect of the crest width L (see Fig. 3.6).

Swamee (1988) gave the following equation of the discharge coefficient for broad-crested weir.

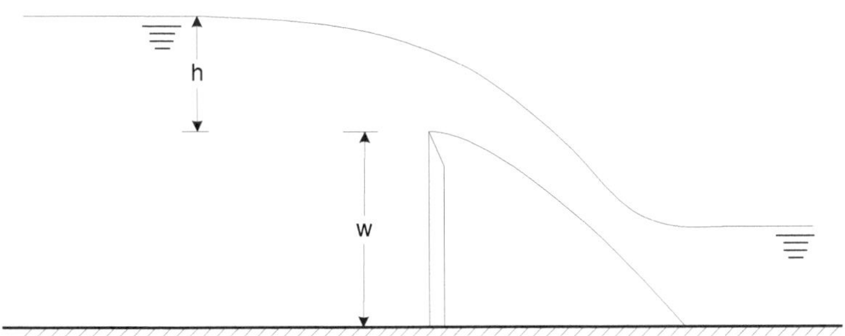

Fig. 3.5 Sharp-crested weir

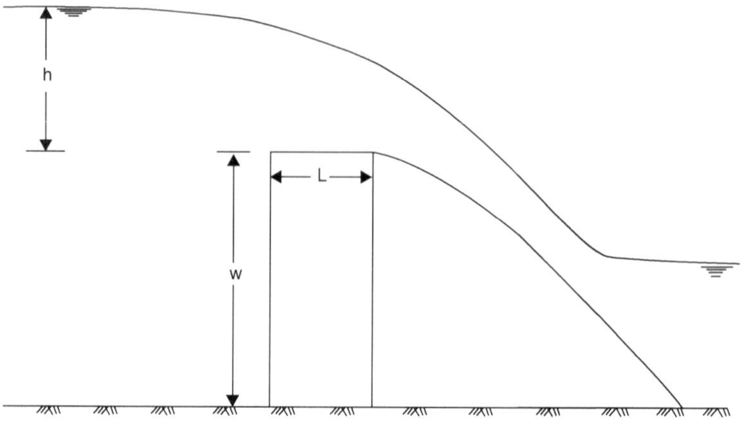

Fig. 3.6 Broad-crested weir

$$C_d = 0.5 + 0.1\left[\frac{\left(\frac{h}{L}\right)^5 + 1,500\left(\frac{h}{L}\right)^{13}}{1 + \left(\frac{h}{L}\right)^3}\right]^{0.1} \tag{3.4.3}$$

Equation (3.4.3) holds good for all values of *h/L* which are less than unity. As *h/L* is increased beyond unity, the weir gradually behaves as a sharp-crested weir, and for $h/L > 1.5$, it is finally converted to a sharp-crested weir. In this transition, both *h/w* and *h/L* govern the discharge coefficient, whereas C_d depends on *h/w* only for the sharp-crested weir and on *h/L* only for a finite-crest weir. Viscosity and surface tension effects are of secondary importance.

3.4.2 Linear Weir

Linear or proportionate weir has greater advantage over the rectangular weir described above. A linear weir is designed to achieve a linear discharge head relationship. For a linear weir, the proportionate error in discharge computation is equal to the proportionate error in head measurement. Thus, the resulting error in the discharge measurement is independent of the head over the weir. On the other hand, there is a large variation of proportionate error in the other shapes of the weir. A linear weir profile proposed by Swamee et al. (1991) has been depicted in Fig. 3.7.

The discharge equation for this weir is

$$Q = 0.166\pi bh\sqrt{gh_*}; \quad h \geq 4h_* \tag{3.4.4}$$

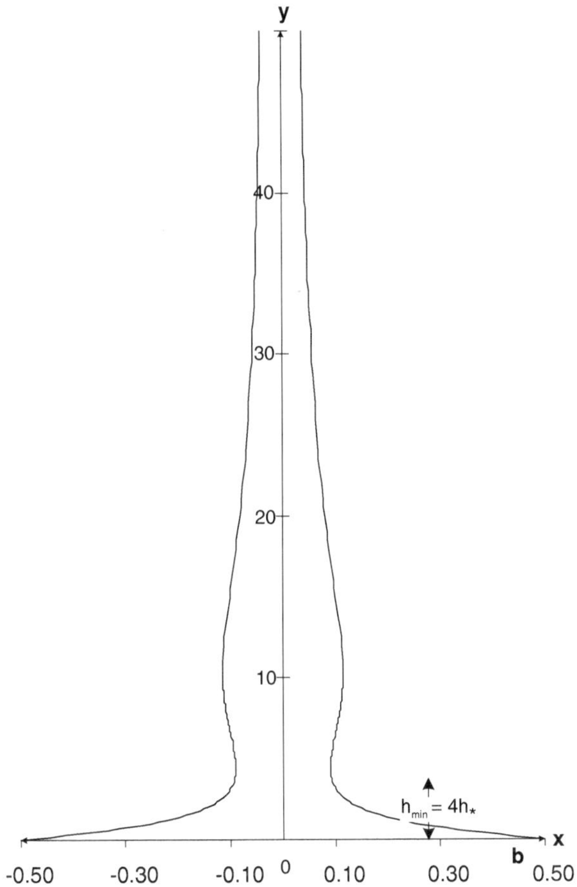

Fig. 3.7 Linear weir profile

where $h_* =$ a length scale that depends on the minimum discharge to be measured. It was recommended by Swamee et al. (1991) that $w = 1.5h_*$. The coordinates of the weir profile are given in Table 3.2.

Example 3.3 Design a linear weir for a rectangular channel having bed width 1.5 m. The minimum and maximum discharges to be measured are 0.05 m^3/s and 0.3 m^3/s, respectively.

Solution Putting $h = 4h_*$ and $Q = 0.0.05$ m^3/s in Eq. (3.4.4) gives $0.05 = 0.166\pi \times 1.5 \times 4.0h_*\sqrt{9.8h_*}$, which yields $h_* = 0.030\,\text{m}$. Since $h_*/w = 1.5$, $w = 1.5 \times 0.03 = 0.045$ m. Similarly, for maximum discharge, Eq. (3.4.4) is written as $0.3 = 0.166\pi \times 1.5h_{\max}\sqrt{9.8 \times 0.03}$, which yields $h_{max} = 0.708$ m. Therefore, for the maximum discharge, the depth of water at the weir is $h_{\max} + w = 0.708 + 0.045 = 0.753\,\text{m}$.

Table 3.2 Linear weir coordinates

h	b	h	b	h	b
0.0	0.5000	6.2	0.0981	22	0.0750
0.2	0.4446	6.4	0.0994	23	0.0721
0.4	0.3854	6.6	0.1008	24	0.0695
0.6	0.3329	6.8	0.1021	25	0.067
0.8	0.2883	7.0	0.1031	26	0.0648
1.0	0.2512	7.2	0.1047	27	0.0627
1.2	0.2205	7.4	0.1057	28	0.0607
1.4	0.1952	7.6	0.1070	29	0.0589
1.6	0.1743	7.8	0.1080	30	0.0573
1.8	0.1571	8.0	0.1090	31	0.0557
2.0	0.1430	8.2	0.1099	32	0.0543
2.2	0.1313	8.4	0.1107	33	0.0529
2.4	0.1218	8.6	0.1115	34	0.0517
2.6	0.1140	8.8	0.1121	35	0.0505
2.8	0.1078	9.0	0.1127	36	0.0494
3.0	0.1027	9.2	0.1131	37	0.0484
3.2	0.0988	9.4	0.1135	38	0.0474
3.4	0.0957	9.6	0.1138	39	0.0465
3.6	0.0934	9.8	0.1140	40	0.0456
3.8	0.0918	10	0.1141	41	0.0448
4.0	0.0907	11	0.1137	42	0.044
4.2	0.0901	12	0.1117	43	0.0433
4.4	0.0899	13	0.1087	44	0.0426
4.6	0.0901	14	0.1050	45	0.0419
4.8	0.0905	15	0.1010	46	0.0413
5.0	0.0912	16	0.0968	47	0.0407
5.2	0.0921	17	0.0927	48	0.0401
5.4	0.0931	18	0.0887	49	0.0396
5.6	0.0942	19	0.0849	50	0.0391
5.8	0.0955	20	0.0814		
6.0	0.0942	21	0.0781		

3.5 Critical Flow

In a canal design, it has to be insured that flow remains subcritical in majority of its length. However, in short reaches, supercritical flow may be allowed. For this purpose, it is necessary to determine critical depth so that if in some reach supercritical flow works out, subcritical flow can be achieved by decreasing the canal bed slope or increasing the canal surface roughness by the way of changing the canal lining. Thus, it is necessary to know critical depth and critical slope.

3.5.1 Critical Depth

Critical flow in an open channel is described by the relationship

$$\frac{Q^2}{g} = \frac{A^3}{T} \tag{3.5.1}$$

3.5.1.1 Power Law Canal Section

A power law section is described by Eq. (2.1.5). Swamee (1993) gave the following equation for critical depth y_c for a power law canal:

$$y_c = \left[\left(\frac{p+1}{p}\right)^3 \frac{Q^2 k_p^2}{4g}\right]^{\frac{p}{3p+2}} \tag{3.5.2}$$

For $p = 2$, a power law section gives a parabolic section. Putting $p = 2$ in Eq. (3.5.2), the critical depth is

$$y_c = \left(\frac{27Q^2}{32ga}\right)^{\frac{1}{4}} \tag{3.5.3}$$

where $a = 1/k_p^2$ is latus rectum of the parabola.

3.5.1.2 Trapezoidal Canal Section

For a trapezoidal channel as shown in Fig. 2.1c, Swamee (1993) gave the following equation for critical depth:

$$y_c = \left[\left(\frac{gb^2}{Q^2}\right)^{0.7} + \left(\frac{gm^2}{2Q^2}\right)^{0.42}\right]^{-0.476} \tag{3.5.4}$$

For $m = 0$, Eq. (3.5.2) reduces to the following critical depth equation or a rectangular channel:

$$y_c = \left(\frac{Q^2}{gb^2}\right)^{\frac{1}{3}} \tag{3.5.5}$$

Similarly for $b = 0$, Eq. (3.5.2) transforms to the critical depth equation for a triangular channel. That is,

$$y_c = \left(\frac{2Q^2}{gm^2}\right)^{\frac{1}{5}} \tag{3.5.6}$$

For a trapezoidal canal having two different side slopes m_1 and m_2, Swamee (1999) generalized Eq. (3.5.4) as

$$y_c = \left\{\left(\frac{gb^2}{Q^2}\right)^{0.7} + \left[\frac{g(m_1+m_2)^2}{8Q^2}\right]^{0.42}\right\}^{-0.476} \tag{3.5.7}$$

3.5.1.3 Circular Canal Section

$$y_c = D\left(0.77\frac{g^3 D^{15}}{Q^6} + 1\right)^{-0.085} \tag{3.5.8}$$

3.5.2 *Critical Slope and Limit Slope*

Knowledge of critical slope is helpful in ensuring that for the stipulated variation of discharge, the canal will have subcritical flow under uniform flow conditions. The critical slope S_c is defined as the slope that would make the given discharge in uniform movement in critical state. Obviously for this slope, the normal depth is equal to the critical depth. For a rectangular channel, S_c exhibits a minimum called the limit slope S_L. For $S_o < S_L$, the channel slope is mild for all discharges. Thus, such a channel will never have supercritical flow under uniform flow conditions. It is, therefore, desirable for a canal to be laid on a slope which is less than the limit slope. For a trapezoidal channel, Swamee (2002) obtained the following equation for limit slope:

$$S_L = 0.0168\,(1-0.3m)\left[5.33\frac{\varepsilon}{b} + \left(\frac{\nu}{b\sqrt{gb}}\right)^{0.51}\right]^{0.27} \tag{3.5.9}$$

Equation (3.5.9) is valid for $0 \le m \le 0.4$. For $m > 0.4$, a limit slope does not exist for any value of ε/b or $\nu/\left(b\sqrt{gb}\right)$.

Exercises

1. A canal carries a discharge of 150 m^3/s at a longitudinal slope 1 in 2000. Assuming representative values of ε and ν, calculate normal depth of flow for the following cases: (a) triangular canal having side slope 1.25, (b) trapezoidal

canal having bed width 15 m and side slope 1.2, (c) rectangular canal having bed width 18 m, and (d) 16 m diameter circular canal.

2. A canal carries a discharge of 50 m^3/s at a longitudinal slope 1 in 200. Assuming suitable values of ε and ν, calculate critical depth of flow for the following cases: (a) triangular canal having side slope 1.5, (b) trapezoidal canal having bed width 5 m and side slope 1.5, (c) rectangular canal having bed width 8 m, and (d) 6 m diameter circular canal.
3. What is the value of limiting slope for a trapezoidal canal having bed width 8 m and side slope 1.4 to carry a discharge of 75 m^3/s?

References

Abramowitz M, Stegun IA (1972) Handbook of mathematical functions with formulas, graphs, and mathematical tables. Dover Publication, New York

Barnes HH Jr (1967) Roughness characteristics of natural channels, Water-supply paper 1849. U.S. Geological Survey, Washington, DC

Chow VT (1973) Open channel hydraulics. McGraw-Hill International Book Co, Singapore

Christensen BA (1984) Discussion on 'flow velocities in pipelines', by Richard R. Pomeroy. J Hydraul Eng ASCE 110(10):1510–1512

Cornish RJ (1928) Flow in a pipe of rectangular cross section. Proc R Soc Lond Ser A 120:691–700

Cunningham AJC (1880) Roorkee hydraulic experiments. Thomason College Press, Roorkee

Cunningham AJC (1882) Recent hydraulic experiments. Proc Inst Civ Eng Lond 71:1–94

Davis SJ, White CM (1928) An experimental study of flow of water in pipes of rectangular section. Proc R Soc Lond Ser A 119:92–107

Friction factors in open channels (1963) Progress report of the task force on friction in open channels of the Committee on Hydromechanics of the Hydraulics Division. J Hydraul Eng ASCE 89(2):97–137

Hager WH (1989) Discussion of 'Non circular sewer design.' by Prabhata K. Swamee, Renu Bhargava, and Ashok K. Sharma. J Environ Eng ASCE 115(1):274–276

Henderson FM (1966) Open channel flow. Macmillan, New York

Henry HR (1950) Discussion of 'diffusion of submerged Jets', by M L. Albertson, Y. B. Dai, R. A. Jensen, and H. Rouse. Trans ASCE 115:687–694

Mital KV (1986) History of Thomason college of engineering. University of Roorkee, Roorkee

Rajaratnam N, Subramanya K (1967) Flow equations for the sluice gate. J Irrig Drain Eng ASCE 93(3):167–186

Rouse H (1956) Discussion on 'a note on the manning formula' by Ven Te Chow. Trans AGU 37(3):327–328

Rouse H, Ince S (1963) History of hydraulics. Dover Publication, New York

Stepanoff AJ (1969) Gravity flow of solids and transportation of solids in suspension. Wiley, New York

Straub LG, Silberman E, Nelson HC (1958) Open channel flow at small Reynolds number. Trans ASCE 123:685–714

Swamee PK (1988) Generalized rectangular weir equations. J Hydraul Eng ASCE 114(8):945–949

Swamee PK (1992) Sluice gate discharge equations. J Irrig Drain Eng ASCE 118(1):56–60

Swamee PK (1993) Critical depth equations for irrigation canals. J Irrig Drain Eng ASCE 119(2):400–409

Swamee PK (1994) Normal depth equations for irrigation canals. J Irrig Drain Eng ASCE 120(5):942–948

Swamee PK (1999) Discussion of 'formula for calculating critical depth of trapezoidal open channel', by Zhengzhong Wang. J Hydraul Eng ASCE 125(7):785–786

Swamee PK (2002) Critical slope equations for open channels. ISH J Hydraul Eng 8(2):44–49

Swamee PK, Chahar BR (2009) Normal depth equation for viscous/laminar flow in a rectangular channel section. ISH J Hydraul Eng 15(2):81–88

Swamee PK, Chahar BR (2010) Normal depth in natural channel sections. ISH J Hydraul Eng 16(1):132–147

Swamee PK, Rathie PN (2012) Normal depth equations for wide rectangular and triangular open-channel sections involving Lambert's W function. ISH J Hydraul Eng 18(3):252–257. Taylor and Francis

Swamee PK, Swamee N (2004) Design of sediment-transporting canal sections. Int J Sediment Res 19(4):312–318

Swamee PK, Swamee N (2008) Design of noncircular sewers sections. J Hydraul Res IAHR 46(2):277–281

Swamee PK, Pathak SK, Agarwal M, Ansari AS (1991) An alternate linear weir design. J Irrig Drain Eng ASCE 117(3):311–323

Swamee PK, Pathak SK, Ali MS (1993) Rectangular side sluice gate analysis. J Irrig Drain Eng ASCE 119(6):1026–1035

Swamee PK, Pathak SK, Ali MS (1994) Side weir analysis using elementary discharge coefficient. J Irrig Drain Eng ASCE 120(4):742–755

Vanoni VA (1975) Sedimentation engineering. ASCE, New York

Williams GP (1970) Manning formula- a misnomer? J Hydraul Eng ASCE 96(1):193–201

Woener JL, Jones BA Jr, Fenzl RN (1968) Laminar flow in finitely wide rectangular channels. J Hydraul Eng ASCE 94(3):691–704

Woo DC, Brater EF (1961) Laminar flow in rough rectangular channels. Geophys Res AGU 66(12):4207–4217

Chapter 4
General Principles of Canal Design

Abstract In design of canals, various factors are considered, for example, the kind of material forming the channel surface which determines the roughness coefficient; the minimum permissible velocity to avoid deposition of silt or debris; the limiting velocity to avoid erosion of the channel surface; and the topography of the channel route which fixes the channel bed slope. This chapter describes general principles of canal design, which includes essential input parameters, and safety and system constraints. Canal discharge is the most important parameter in designing a canal. The chapter presents how to determine design discharge for irrigation canals and power canals. Also, the detailed information on canal lining and material selection, drainage and pressure relief arrangements, requirements on canal banks and freeboard, longitudinal bed slope, canal section shape selection, typical canal layout, and necessity of cross-drainage works is included in this chapter.

Keywords Safety constraints • System constraints • Design discharge • Command area • Duty • Power canal • Dependable flow • Lining material • Pressure relief • Drainage • Uplift pressure • Canal banks • Freeboard • Longitudinal bed slope • Canal layout • Cross-drainage works • Aqueduct • Superpassage • Siphon • Level crossing • Inlet and outlet

A canal has to be designed in such a way so as to minimize its cost keeping the aim of supplying the requisite quantity of fluid at the prescribed slope. In Chap. 2, cost functions of various components of a canal have been formulated which can be utilized in the design of a canal based on cost considerations. In the present form, disregarding reliability, we restrict ourselves only to the cost of the canal. Thus, minimization of cost is the objective of the design. In such a problem, the cost function is the objective function of a canal. The objective function F is a function of *decision variables* (which are commonly known as *design variables*) like bed width and side slope for a trapezoidal canal. That is,

$$F = F\,(\text{decision variables}) \tag{4.1}$$

P.K. Swamee, B.R. Chahar, *Design of Canals*, Springer Transactions in Civil and Environmental Engineering, DOI 10.1007/978-81-322-2322-1_4

4.1 Constraints

The problem is to minimize the objective function F. By selecting the design variables (bed width, normal depth, etc.) to zero, the objective function can be reduced to zero. This is not an acceptable situation as there will be no canal, and the objective of fluid transport will not be achieved. In order to exclude such a solution, additional condition of transporting the fluid has to be prescribed. In addition to this, constraints of minimum bed width (or diameter) and maximum average velocity have to be observed. The restriction of minimum bed width (or diameter) is from maintenance considerations, whereas the restriction of maximum average velocity avoids excessive velocities that are injurious to the canal lining. In practice, limits may be encountered on canal depth due to existence of unfavorable strata and/or groundwater at a shallow depth and on top width due to restrictions on the span of bridges and cross-drainage works or on the width of right of way. Sometimes the side slope is chosen based on the angle of repose of material for better stability. Higher velocities are desired in rigid boundary channels to reduce costs. A designer may refer to local code of practice for prescribed limiting velocities for different types of linings. These restrictions on canal dimensions or flow velocity are called *safety constraints*.

4.1.1 Safety Constraints

The minimum bed width (and diameter) constraint can be written as

$$b \geq b_{\min} \tag{4.1.1a}$$

$$D \geq D_{\min} \tag{4.1.1b}$$

where $b_{\min}$ and $D_{\min}$ = the minimum prescribed bed width and diameter, respectively. For values of $b_{\min}$ and $D_{\min}$, local code of practice may be consulted.

The minimum velocity constraint is written as

$$\frac{Q}{A} \geq V_{\min} \tag{4.1.2}$$

where $V_{\min}$ = minimum permissible velocity or non-silting velocity, which is the lowest velocity that will not initiate sedimentation and will not allow the growth of vegetation. Sedimentation and growth of vegetation decrease the carrying capacity and increase the maintenance cost of a canal. In general, an average velocity of 0.6 to 0.9 m/s will prevent sedimentation when the silt load of the flow is low, and a velocity of 0.75 m/s is usually sufficient to prevent the growth of vegetation (Chow

Table 4.1 Limiting velocities

Lining material	Limiting velocity (m/s)
(1)	(2)
Sandy soil	0.3–0.6
Black cotton soil	0.6–0.9
Muram and hard soil	0.9–1.1
Firm clay and loam	0.9–1.15
Gravel	1.0–1.25
Disintegrated rock	1.3–1.5
Boulder	1.0–1.5
Brunt clay tile	1.5–2.0
Concrete tile	2.0–2.5
Concrete	2.5–3.0
Hard rock	3.0–4.0

1973). Hence, the minimum permissible velocity can be assumed in the range from 0.75 to 0.9 m/s. The maximum velocity constraint is written as

$$\frac{Q}{A} \leq V_L \tag{4.1.3}$$

where $V_L =$ maximum allowable velocity or limiting velocity. The limiting velocity depends on the canal lining material. Table 4.1 (Sharma and Chawla 1975; Subramanya 2009) lists the limiting velocity for various lining materials. If V is greater than V_L, superior lining material having larger V_L is selected.

4.1.2 *System Constraint*

The designed canal must convey the requisite discharge Q. This implies that the design variables must satisfy the resistance equation for the given discharge. The resistance equation for flow of a viscous fluid in a rectangular channel is given by Eq. (3.1.5), whereas for a turbulent flow, it is given by Eq. (3.1.15). On the other hand, the resistance equation for flow of a sediment transporting canal is given by Eq. (3.1.22).

4.2 Formulation of the Problem

The canal design problem, thus, boils down to minimization of Eq. (4.1) subject to the constraints given by Eq. (3.1.5) or (3.1.15) or (3.1.22) as the case may be. The objective function is nonlinear in the design variables. Similarly, maximum velocity constraints given by Eq. (4.1.3) is a nonlinear inequation. Such a problem is hard to be solved mathematically. However, it can be routinely solved numerically. Many numerical algorithms have been devised from time to time to solve such problems.

The calculated bed width (or diameter) and side slope cannot be provided in the actual practice. Thus, the designer has to select a rounded off size. The rounding off is generally carried by rounding up. Similarly, if the number of canal falls may involve a fractional part, the next higher number has to be adopted for the number of falls.

4.3 Essential Parameters for Canal Design

The selection of the design discharge for irrigation canals and power canals, freeboard, and reliability considerations are some of the important parameters required to be selected before designing any water system. The bed slope of a canal is normally decided at an early stage of design. It can be made as flat as possible to keep command or steeper to reduce canal size or increase velocity. A brief description of these parameters is provided in this section.

4.3.1 Canal Discharge

Canal discharge or capacity is the most important parameter to be prescribed for designing a canal. It is also necessary in managing the canal system. For irrigation canals, the water requirement depends upon the command area, the type of cropping pattern and irrigation practices, and the rainfall on the command area. *Gross command area* is defined as total area that can be irrigated by a canal system on the perception that unlimited quantity of water is available. It is the total area that may theoretically be irrigated by the canal system. But this may include inhibited areas, roads, ponds, uncultivable areas, etc. *Culturable command area* is the actual irrigated area and is fit for cultivation. However, the entire culturable command area is never put under cultivation during any crop season. *Intensity of irrigation* is defined as the proportion of the culturable command area to be irrigated annually. Usually the area irrigated during each crop season is expressed as a proportion of the culturable command area, which represents the intensity of irrigation for the crop season. By adding the intensities of irrigation for all crop seasons, the yearly intensity of irrigation is obtained. The intensity of irrigation may vary from 0.4 to 1.8. The culturable command area multiplied by the intensity of irrigation gives the actual area to be irrigated.

The *duty* is a measure of design capacity of the canal. It is expressed either as discharge required to irrigate unit *command area* or as an area that can be irrigated per unit discharge. Here the command area is the land that can physically be irrigated by a canal system. The crop water requirement depends upon climatic factors like rainfall pattern, temperature, wind movement, and relative humidity as well as upon soil type in relation to water holding and infiltration, position of subsoil water table, slope of the ground, drainage condition of the area, and host of other factors. The field water demand is calculated as the water requirement of the command area for assumed cropping pattern and intensity of irrigation. The duty varies along the system being maximum at the headworks and minimum at the farm due to various losses. Not all the water that diverted at canal headworks gets to where it is intended. Water always gets lost along the canal way. It may be lost as leakage, seepage, evaporation, overtopping, dead storage, bad management, theft/wastage by human/animals, etc. Seepage from the canal can be heavy if the soil is permeable and canal is unlined or poorly lined. There are many empirical methods and rules of thumb for estimation of seepage loss from canals. Theoretical methods for estimation of seepage loss are dealt in Chap. 2. The overall efficiency of a canal system is a measure of the total losses. It is the amount of water needed by the crop divided by the actual amount of water diverted at canal headworks. In case the water availability at the river intake is less than the crop demand, then the river water availability has a canal-carrying capacity. Normally, 75 % of the total available water may be diverted for irrigation after priority requirements like drinking water supply, industrial water supply, etc., are met.

Example 4.1 The gross command area of a canal is 100,000 ha, 80 % of which can be irrigated. The intensities of irrigation for nonoverlapping crop seasons are 60 % and 30 % and for overlapping crop season is 10 %. Determine the required discharge at canal head, if the average duty at the canal head is 2,000 ha per m^3/s for crops in the first nonoverlapping season, 1,200 ha per m^3/s for crops in the second nonoverlapping season, and 800 ha per m^3/s for crops in the overlapping season. Assume canal losses as 25 %.

Solution Cultural command area = $0.80 \times 100{,}000 = 80{,}000$ ha. Areas under irrigation in nonoverlapping crop seasons are $0.60 \times 80{,}000 = 48{,}000$ ha and $0.30 \times 80{,}000 = 24{,}000$ ha and in overlapping crop season $0.10 \times 80{,}000 = 8{,}000$ ha. Discharges required for nonoverlapping crop seasons are $48{,}000/2{,}000 = 24\ m^3/s$ and $24{,}000/1{,}200 = 20\ m^3/s$ and for overlapping crop season $8{,}000/800 = 10\ m^3/s$; thus, the total crop water requirement at head is max $(24 + 10, 20 + 10) = 34\ m^3/s$. Therefore, the design discharge considering 25 % conveyance losses is $34/0.75 = 45.33\ m^3/s$.

Canal in hydropower plants takes off from the head regulator (upstream of the plant) and meets the stream again on downstream of the power plant. Canals in hydropower plants are of three types: (1) feeder canal to connect the head regulator (intake) to the desilting basins, (2) power canal to connect the desilting basins to the forebay tank, and (3) tailrace canal to connect the draft-tube/turbine chamber to the river. See Fig. 4.1.

Feeder canal is designed for turbine flow plus desilting flushing flow. A sufficiently high velocity must be provided to prevent deposition of sediment within the canal. Power canal is designed for total turbine flow. A freeboard allowance above

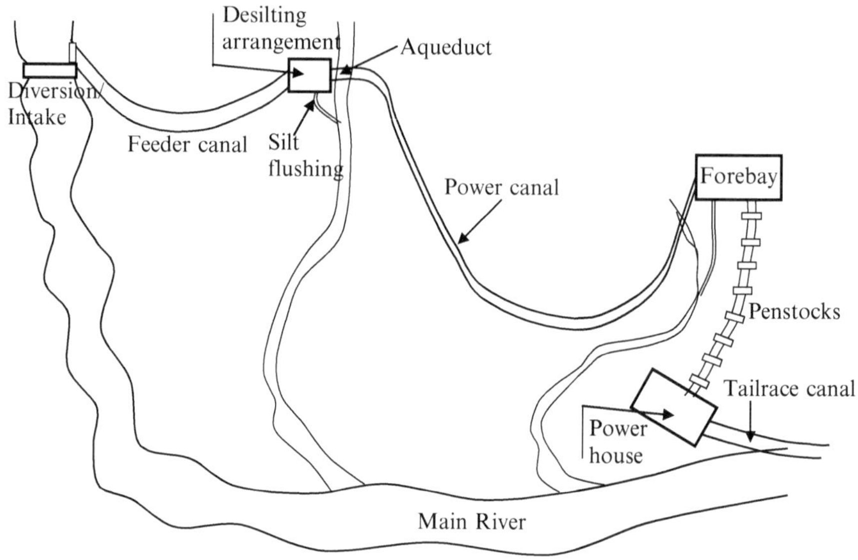

Fig. 4.1 Layout of feeder canal, power canal, and tailrace canal in a hydropower plant

the steady-state design level is required to contain water safety within the canal in event of power outages. After passing through the turbine, water goes to the river via the tailrace canal. The tailrace canal is oriented to discharge at an angle of 30°–45° to the center line of the receiving river. This will help in keeping the tailrace channel clear of bed load deposits that could cause back water effects at the powerhouse or require expensive maintenance dredging to control water levels. Power canals, also called *Hydel* canals, are designed for 90 % of annual dependable flow (Arora 2007). 90 % dependable flow is the flow which is likely to be exceeded in 90 % duration of the total data period. During planning of hydropower projects, discharge data of the river for about 30 continuous years is taken into consideration. Annual discharge data are arranged in descending order and exceedance probability is computed. Based on the exceedance probability, 90 %, 75 %, and 50 % dependable years are identified. If discharge data for N years are available and are arranged in descending order in a table, the 90 % dependable year is defined as $(N + 1) \times 0.9$ year in the table. Water availability is usually computed using 10-daily flows. The average of 10 consecutive daily flows is known as 10-daily flows. Usually, 10-daily flows are computed on a monthly basis as three sequences, i.e., I, 1 to 10 days of the month; II, 11 to 20 days of the month; and III, 21 to 30 days of the month. For different dependable flows, power potential studies based on 10-daily flows are carried out to fix the design discharge most appropriate for the installed capacity. This procedure is applicable if there is no restriction on the use of river water for power generation. For restricted water use case, the design discharge is the maximum surplus water allowed for power generation. See Table 4.2 for dependable flow calculation for discharge data of 29 years from 1970 to 1998 in a tributary of the Beas River. It can

Table 4.2 Computation of dependable flows

Rank i	Year	Annual runoff (10^6 m^3)	Exceedance probability $\frac{i}{N+1} \times 100$
1	1973	365.0	3.33
2	1988	265.4	6.67
3	1987	245.4	10.00
4	1986	242.5	13.33
5	1976	224.3	16.67
6	1983	222.1	20.00
7	1990	214.5	23.33
8	1977	211.9	26.67
9	1974	207.9	30.00
10	1975	207.1	33.33
11	1994	205.8	36.67
12	1982	202.6	40.00
13	1998	200.6	43.33
14	1992	187.2	46.67
15	**1993**	**178.0**	**50.00**
16	1995	177.7	53.33
17	1996	174.9	56.67
18	1985	169.5	60.00
19	1991	166.9	63.33
20	1989	162.6	66.67
21	1978	161.8	70.00
22	1981	147.0	73.33
23	**1971**	**142.8**	**76.67**
24	1980	142.7	80.00
25	1972	137.9	83.33
26	1997	134.1	86.67
27	**1984**	**122.3**	**90.00**
28	1970	104.1	93.33
29	1979	85.3	96.67

$N = 29$ (total number of data years)

be seen that 90 %, 75 %, and 50 % dependable flows occur in 1984, 1971, and 1993, respectively. In 1993, 50 % dependable flow = 5.65 m^3/s (178.0 × 10^6 m^3/year); in 1971, 75 % dependable flow = 4.53 m^3/s (142.8 × 10^6 m^3/year); and in 1984, 90 % dependable flow = 3.88 m^3/s (122.3 × 10^6 m^3/year).

Table 4.3 lists 10-daily flow sequences for 90 %, 75 %, and 50 % dependable flow years to be used for power potential calculations. With 90 % dependable flow criteria, the design discharge for power canal would be 3.88 m^3/s. Alternatively, the design discharge may be fixed based on 75 % dependable flow or power potential studies.

Table 4.3 Design discharge for a power canal

S. no	Dependable discharge (m^3/s)				
	Month	10-daily sequence	50 % (1993)	75 % (1971)	90 % (1984)
1.	January	I	2.76	0.99	2.13
2.	January	II	2.64	0.95	2.15
3.	January	III	2.41	0.72	1.86
4.	February	I	2.58	1.07	1.80
5.	February	II	2.69	1.11	2.00
6.	February	III	2.50	1.10	1.91
7.	March	I	2.62	1.41	2.16
8.	March	II	2.73	1.81	2.03
9.	March	III	3.40	2.69	2.50
10.	April	I	2.85	2.69	2.70
11.	April	II	3.66	3.46	2.69
12.	April	III	5.45	3.84	2.70
13.	May	I	7.64	1.47	2.84
14.	May	II	5.18	3.64	2.70
15.	May	III	8.48	4.98	4.68
16.	June	I	7.66	7.57	4.98
17.	June	II	9.30	6.51	5.34
18.	June	III	9.59	7.39	4.57
19.	July	I	12.08	13.55	6.18
20.	July	II	23.82	9.74	6.98
21.	July	III	11.00	17.10	9.84
22.	August	I	10.52	14.95	10.18
23.	August	II	8.28	12.03	8.96
24.	August	III	7.08	9.35	8.86
25.	September	I	8.26	9.77	10.12
26.	September	II	8.55	5.25	5.87
27.	September	III	5.64	3.41	3.22
28.	October	I	4.60	2.78	2.66
29.	October	II	3.94	2.58	2.51
30.	October	III	3.27	1.81	2.22
31.	November	I	2.79	1.64	2.11
32.	November	II	2.38	1.41	1.94
33.	November	III	2.03	1.20	1.64
34.	December	I	1.80	1.08	1.50
35.	December	II	1.60	0.99	1.59
36.	December	III	1.46	0.89	1.47

4.3.2 Canal Lining and Material Selection

Canals are lined for several reasons. As the lined canals permit higher average velocities, there is a saving in the cross-sectional area of the canal and land acquisition with corresponding saving in the cost of excavation and masonry works. Further, a lined canal can be laid on steep slopes to save the cost of earthwork in formation. Furthermore, the smooth surface of lining reduces the friction loss, which enables the canal to be laid on a flatter bed slope. A flatter bed slope increases command area and provides larger working head for power generation. The maintenance cost of a lined canal gets reduced as the lining ensures protection against bed and bank erosion. The canal lining becomes essential for stability of the canal banks in expansive soils. Canals conveying water in arid and semiarid regions need to be lined prior to conveying water in them. This is because, for a newly completed irrigation project, seepage from canal is more as the groundwater level is likely to be at a larger depth below the canal bed. Also in arid and semiarid regions, the groundwater is likely to be brackish and the seepage water which joins the groundwater may not be withdrawn by pumping as the pumped water is unlikely to satisfy irrigation water standards. Control of seepage saves water for further extension of the irrigation network and reduces the water logging in the adjoining areas. Seepage control for existing canals with conventional lining is found to be prohibitive because of the high cost of material and restriction on closure of canal.

Several types of materials are used for canal lining. The choice of material mainly depends on the degree of watertightness required along with other factors such as economy, structural stability, strength and durability, reparability and easy maintenance, hydraulic efficiency, resistance to erosion, ability to prevent weed growth, resistance against burrowing animals, and flexibility. Soil-cement lining and boulder lining are preferred on account of their low initial cost, though they are less watertight. Other low-cost linings are polyethylene plastic, alkathene sheets, asphalt membranes, and bentonite clay membranes, which are spread over the boundary surface with adequate earth cover. These types of lining are used for stable channels. Brick lining and brunt clay tile lining are popular linings as they provide reasonable watertightness along with strength. For improving these qualities, double layers of brick or brunt clay tiles are used. For canals carrying large discharges, in situ concrete lining or concrete tile linings are used. Low-density polyethylene (LDPE) or high-density polyethylene (HDPE) films of thickness ranging from 100 to 200 μm or more are sandwiched between two layers of canal linings to prevent seepage.

4.3.2.1 Roughness Height

General resistance equation involves physical conceivable parameter ε (average roughness height of the channel surface), which depends on the finished surface of the lining. The smoother is the surface, the lesser will be the roughness height and consequent resistance to flow or head loss. In practice, the roughness height of a finished surface of the lining can be estimated by a hydraulic method. The method comprises the measurement of flow depth and corresponding discharge in a fairly long channel section. In a long cannel, the flow becomes uniform at normal depth. The flow depth can be measured with a point gauge or other suitable arrangements, while the discharge can be measured with a volumetric method, weir arrangement, ADV/LDV, or other flow measuring arrangements. For a fixed channel, bed width, side slope, and bed slope are already known. Alternatively, a tilting bed flume can be used which has flexibility of varying bed slopes. The kinematic viscosity can be adopted from tables corresponding to the measured temperature of flowing water in the channel. With all these parameters, the resistance equation can be solved for unknown roughness height of the channel surface. The experimental channel can be lined with the material with similar finishing to be used in the actual field canal.

4.3.2.2 Pressure Relief Valves

Canal lining is a thin structure. Unbalanced water pressure on either side of the lining can cause damage. Lining is capable of resisting small differential heads, for example, concrete lining of 100 mm thick can withstand a hydrostatic pressure of about 1 m head. There are mainly two situations in which external back water pressure may develop, namely, high groundwater level and rapid/sudden drawdown (or canal is emptied) in saturated subgrade. If this hydrostatic pressure is excessive, the lining cannot counterbalance it. The drainage arrangements to be provided would depend mainly upon the position of the water table and the type of subgrade. The water table may have the following positions: (1) below the canal bed level, (2) between the canal bed and full supply level, and (3) above the canal full supply level. The subgrade may be of the following types: (1) free draining, (2) poor draining, and (3) practically impervious. The hydrostatic/uplift pressure depends on the relative positions of the water table in the subgrade and the water level in the canal. Excessive hydrostatic pressure sufficient to damage the lining may develop when the water level in the canal is very low (or the canal is empty) and water table is high. The accumulation of water in the poorly draining soil surrounding the canal results in localized high water table, which produces damaging hydrostatic back pressure during a period of rapid drawdown of water level in the canal. Drainage arrangements are provided such that the pressure on the lining does not increase beyond the safe limit. In case of water table above the canal bed level, the lining is subjected to uplift pressure equal to the head difference between the water table and water level in the canal irrespective of the type of subgrade. In case of sudden drawdown, the uplift pressure depends upon the rate of drawdown and the drainage characteristics of the subgrade. In such cases, either the rate of drawdown is controlled or subgrade drainage is provided. Pressure lag in a poorly draining subgrade is also important from the point of view of the stability of the bank itself. For inhibiting the movement of fine material from the subgrade, filter is provided. This prevents formation of cavities which subsequently leads to subsidence of the lining. Therefore, provision of pressure relief valves and underdrainage arrangement becomes essential. In case a lined canal crosses areas subject to seasonal high water table or where the soil is poorly draining (or practically impervious), suitable underdrainage is provided to protect the lining.

The water accumulated by means of the underdrainage arrangements used is to be disposed of by means of natural drainage or by the use of pressure release valves into the canal. Pressure relief valve is a device that allows water behind a canal lining to flow inside the canal but does not allow canal water to leak in the subgrade. These valves, thus, protect the lining from damage by releasing excessive hydrostatic pressure behind the lining. Drainage and pressure release arrangements are not necessary when water table is below the canal bed level and also when the subgrade is freely draining. Drainage and pressure relief arrangements along bed and sides of the canal are required for all three types of subgrade conditions when water table is above the canal bed level. When the subgrade is poorly draining, a 150 to 200 mm thick layer of filter is required. When the subgrade is practically impervious, it necessitates removal of the subgrade up to some depth and replacement by a pervious material.

4.3.3 Canal Banks and Freeboard

4.3.3.1 Canal Banks

The purpose of the bank is to clasp water. Bank widths are determined by local design rules. In case of distributaries, service road is provided on one bank for inspection and maintenance purpose. However, in case of main and branch canals, service road is provided on both the banks. Adequate drainage is provided with suitable slope away from the canal side so that no rainwater flows or percolates toward the canal slope behind the linings. A *dowla* is provided on side of the canal service road as a safety measure in driving (see Fig. 4.2a). It prevents erosion of the slope due to rains. It is provided on one or both the banks depending upon the type and size of the lined canal. In high embankments, the *dowlas* may be replaced by parapets. The minimum values recommended for top width of the bank are as in Table 4.4.

Where the stability of the embankment is required, wider bank widths can be provided. In hilly terrain where it is not possible to provide above bank widths, the bank widths may be suitably reduced.

4.3.3.2 Freeboard

The freeboard is an extra height provided over and above the normal depth up to the top of the lining (lining freeboard) or up to the top of the canal bank (bank freeboard). See Fig. 4.3. It is a safety margin which can accommodate waves,

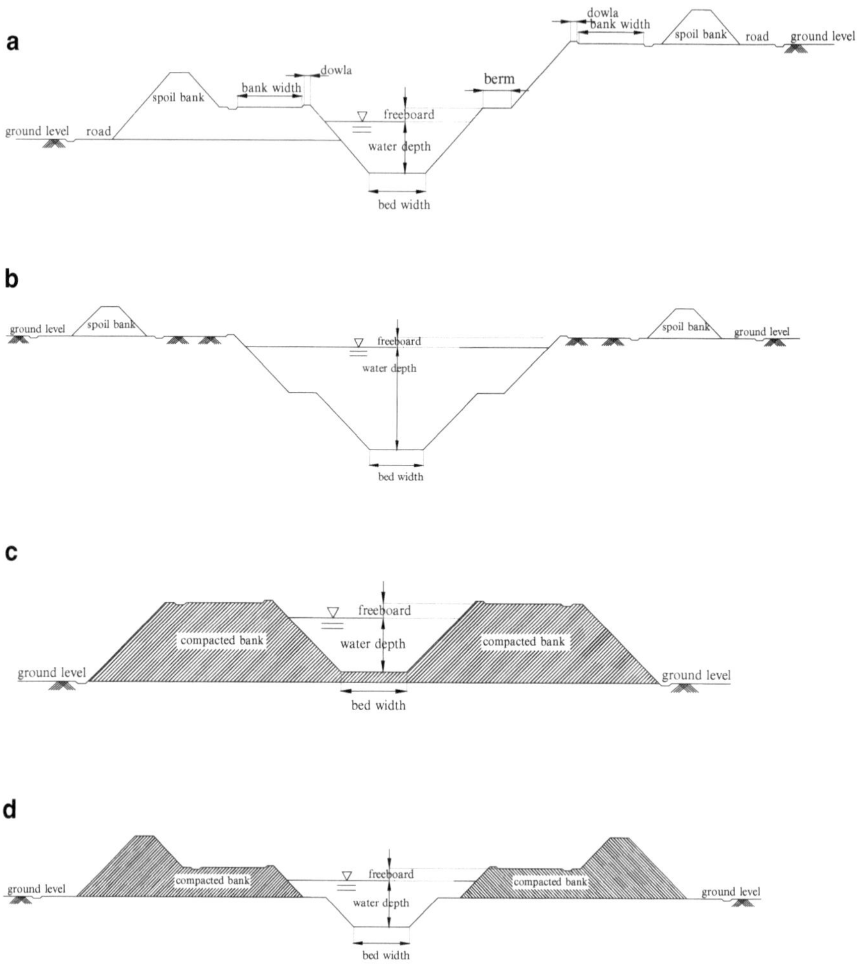

Fig. 4.2 Typical canal sections. (**a**) Canal section with bank and *dowla*. (**b**) Canal in full cutting. (**c**) Canal in full filling. (**d**) Canal in partial cutting and partial filling

Table 4.4 Typical values for banks

Discharge (m^3/s)	Minimum bank top width	
	Inspection bank/wider bank (m)	Non-inspection bank/other banks (m)
0.15–1.5	4.0	1.5
1.5–3.0	4.0	2.0
3.0–10.0	4.0 + *dowla*	2.5
10.0–30.0	5.0 + *dowla*	4.0
30.0 and above	6.0 + *dowla*	5.0

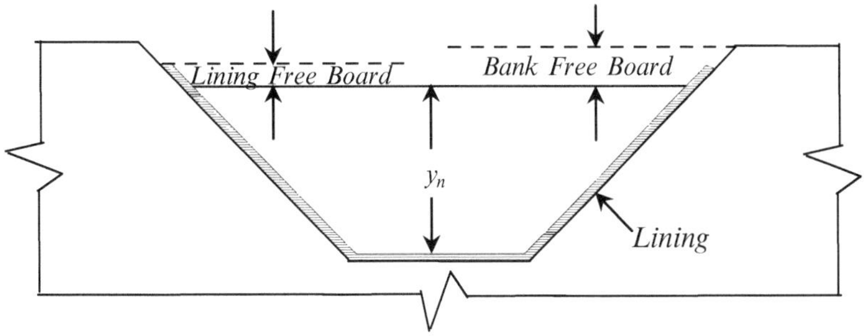

Fig. 4.3 Definition of freeboard

a flood surcharge, or a surge flow caused by faulty operation of the canal or which can absorb effects of poor construction tolerances or inaccurate estimates in roughness/slope.

The freeboard depends upon a number of factors, such as size of channel, velocity of flow, depth of flow, curvature of alignment, condition of storm and drain-water inflow, fluctuations in water level due to operation of flow-regulating structures, and wind action. There is no universally accepted rule for the determination of freeboard, since wave action or water surface fluctuation in a channel may be created by many uncontrolled causes. Pronounced waves and fluctuation of water surfaces are generally expected in channels where the velocity is so high and the slope is so steep that the flow becomes very unstable or on curves where high velocity and large deflection angle may cause appreciable superelevated water surface on the convex side of a curve or in channels where velocity of flow approaches the critical state at which the water may flow at alternate depths and thus jump from the low stage to the high stage at the least obstruction. There are many rules of thumb for determining the height of freeboard.

Discharge-Dependent Freeboard: The discharge-dependent freeboard is constant once the design discharge is known. Various codes of practice recommend different discharge-dependent values varying between 0.1 m and 0.75 m.

Depth-Dependent Freeboard: In opposition to discharge-dependent freeboard, one may also consider depth-dependent freeboard, which will be a function of normal depth y_n. The freeboard based on discharge alone is not quite satisfactory because it remains the same for shallow and deep channels for a given discharge. A depth-dependent freeboard is more general as it accounts the design discharge through the normal depth and also differentiates between shallow and deep channels. A recommended form of depth-dependent board is (Chow 1973)

$$F = \Delta y^{\delta} \tag{4.3.1}$$

where $F =$ freeboard in m, $\delta = 0.5$, and Δ ranges from 0.6735 for $Q = 0.57$ m^3/s to 0.8723 for $Q \geq 85$ m^3/s. However, there seems to be lack of information on the variation of Δ between these limits (Loganathan 1991). The variation of Δ between

Table 4.5 Longitudinal bed slope for different discharges

Discharge (m^3/s)	0.025–0.05	0.1–3.0	2.0–10.0	>10.0
Bed slope	0.005	0.0005	0.0002	0.0001

these limits can be assumed to be linear. To ascertain the value of Δ for different values of Q between above limits, the following equation could be used (Chahar et al. 2007):

$$\Delta = 0.672 + 2.357 \times 10^{-3} Q \tag{4.3.2}$$

4.3.4 Longitudinal Slope of Canal

Longitudinal bed slope, hydraulic slope, hydraulic gradient, and energy line are parallel and the same in canals. Longitudinal bed slope is the elevation difference between two points on the bed along the canal expressed as vertical fall in meters per meter length of canal. It can be made as flat as possible to keep command or steeper to increase velocity and reduce canal size. Mainly it is governed by the general terrain of the area through which the canal is passing. The discharge is also a controlling parameter for the bed slope as the bed slope is kept flatter for higher discharge and steeper for small discharge. The common range of bed slope is listed in Table 4.5. See Laycock (2007).

4.3.5 Canal Section Shape

Canal section can be of any shape. Natural channels often have very complicated shape, which cannot be expressed in terms of mathematical formulae. An unlined canal will in time create its own profile depending on the soil type it passes through and the nature and load of silt it carries. For artificial/manmade channels, the shape is chosen such that it is easy to construct and requires least cost. These criteria limit the choice to a few standard sections. Most canals are designed to a standard trapezoidal cross section. Triangular-shaped canals are preferred for small discharge, machine-graded ditches, grassed waterways, gutters, etc. For carrying large discharges, a rectangular section is not preferred. This is on account of the stability of side slopes. Vertical sidewalls require large thickness to resist the earth pressure. On the other hand, sloping sidewalls of a triangular/trapezoidal section require less thickness. Rectangular sections are commonly used in urban areas, aqueducts (canals above drains or low-lying areas), hard soils, etc. Rectangular shape requires minimum land acquisition and is easy to construct over the piers. Hard soils are easy to cut/blast in rectangular shape. Also, it is easy to construct vertical sides with brick or masonry, but it is difficult to control leakage through junction at base and sidewalls, and vertical sides are weak against horizontal forces from roots, swelling soils, earth pressure, and surcharge (traffic, structure, etc.). Rectangular shape is better for small lined canals used as field channels (or water courses) in farms as the above mentioned negative points are unimportant. For large discharges, trapezoidal shape is universally used because most engineers prefer to work with straight lines rather than curves. Parabolic is for many situations the best practical shape for a variety of reasons (Chahar 2005). It is the best shape to use for small discharge. It retains a higher velocity at low discharge and hence has less tendency to deposit sediment and high chances to carry floating debris. Precast blocks of parabolic shape provide more structural strength, low overall stress, and no stress concentration points. For small discharges, semicircular sections are often adopted for irrigation canals. Semicircle is the best section of all in terms of minimum area and perimeter. But this only applies to a theoretical case with no freeboard. Excavation and in situ lining of large circular section are not easy. For large discharge, compound section is more useful. Compound channel shapes are composed of more than one standard shape. Curved bed/corner trapezoidal section used especially for large canals is a compromise between better properties of curved sections and popular trapezoidal section (Chahar and Basu 2009). Stress concentration problems of the trapezoidal shape are largely reduced in such section. There are several other shapes occasionally used for small canals.

4.3.6 Canal Layout

A canal becomes smaller in size as its capacity reduces from source (headworks) to destination (farm). The starting canal (feeder canal, transmission canal, or main canal) takes off from headworks and may transfer water from surplus river basin to deficit basin. The last canal (water course or field/farm channel) feeds group of farms or individual fields. In between, there may be branch canal, distributary, major, minor, etc. A transmission/feeder canal conveys water from the source to a distribution canal. There is no withdrawal from a transmission canal. Many times, the area to be irrigated lies very far from the source, requiring long transmission canals [e.g., the Indira Gandhi Canal system has a transmission canal length of 204 km and carries a discharge of about 524 m^3/s (Hooja 1993)]. The layout of a canal system needs careful consideration according to topography between the source and destination. Several alignments between the source and the destination are possible. The topography is generally irregular having a network of natural drains and ridges along with local depressions and peaks. A canal is aligned in such a way that it covers the entire area proposed to be irrigated with the shortest possible length and minimum number of cross-drainage works. A shorter length of canal ensures less loss of head and discharge leading to more command area. A shortest distance alignment may not be feasible due to many reasons. Canals are aligned as far as possible in partial cutting and partial filling. Deep cutting or high embankments are generally avoided by suitable detouring. Carrying of a canal in high embankment involves high risk of breaching as well as seepage loss. Thus, a canal alignment problem is a combination of horizontal and vertical layout designs. The number of kinks and acute curves is kept minimum. Radii of curves may be three to seven times the water surface width.

Canal alignment may be contour canal, side slope canal, or ridge canal as per the topography. A contour canal irrigates only one side of the canal and it crosses a number of valleys. A side slope canal is aligned at right angles to the contours of a country. A watershed or ridge canal irrigates the areas on both sides. The main canal is carried on a contour alignment, till it reaches the top of a watershed to become a watershed canal. Branch canals and distributaries take off from a canal from or near the points where the canal crosses watershed (Fig. 4.4). The topography at farm level may be modified to some extent by earthmoving and land leveling. The steepest possible alignment of a canal will cross the contours at right angles.

A canal must negotiate obstacles such as depressions, valleys, rivers, steep terrain, etc., between two points. A cross-drainage work is a structure built on a canal where it crosses a natural drainage, such as a stream or a river. Sometimes, a cross-drainage work is required when the canal crosses another canal. The cross-drainage work is required to dispose of the drainage water so that the canal supply remains uninterrupted. The canals are, preferably, aligned on the watershed so that there are no drainage crossings. However, it is not possible to avoid the drainages in the initial reach of a main canal because it takes off from headworks located on a river which is a valley. The canal, therefore, requires a certain distance before it can mount the watershed (or ridge). In this initial reach, the canal is usually a contour canal, and it intercepts a number of natural drainages flowing from the watershed to the river. After the canal has mounted the watershed, no cross-drainage work will normally be required. A cross-drainage work is an expensive structure and it is avoided as far as possible. The number of cross-drainage works can be reduced to some extent by changing the alignment of the canal. Sometimes it is possible to reduce the number of cross-drainage works by diverting the small drainages into large drainages.

Depending upon the relative positions of the canal and the drainage, the cross-drainage works may be broadly classified as (1) canal over the drainage, aqueduct and siphon aqueduct; (2) canal below the drainage, superpassage and canal siphon; and (3) canal at the same level as the drainage – level crossing, inlet, and inlet and outlet. See Fig. 4.5.

The relative levels and discharges of the canal and of the drainage mainly affect the type of cross-drainage work required. (1) If the canal bed level is sufficiently above the High Flood Level of the drainage, an aqueduct is selected. (2) If the Full Supply Level of the canal is sufficiently below the bed level of the drainage, a superpassage is provided. (3) If the canal bed level is only slightly below the HFL of the drainage, and the drainage is small, a siphon aqueduct is provided. If necessary, the drainage bed is depressed below the canal. (4) If the FSL of the canal is slightly above the bed level of the drainage and the canal is of small size, a canal siphon is provided. (5) If the canal bed and the drainage bed are almost at the same level, a level crossing is provided when the discharge in the drainage is large for a short

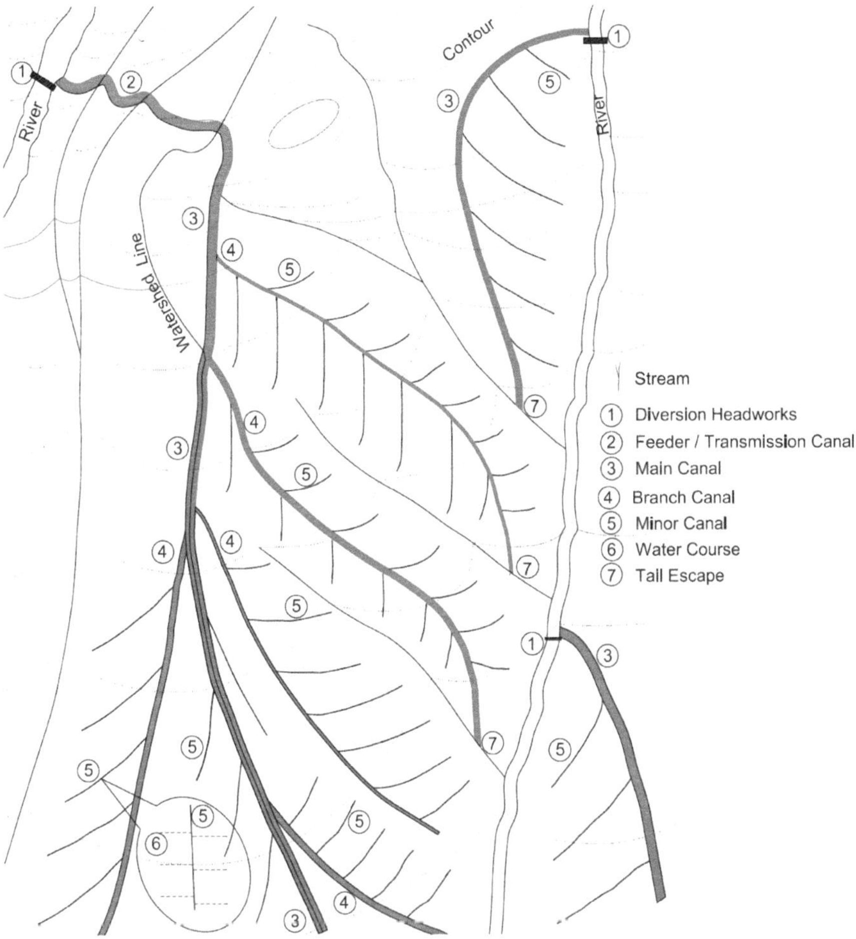

Fig. 4.4 Canal layout

duration. In a level crossing, the drainage water is admitted into the canal at one bank and is taken out at the opposite bank. An inlet alone may be provided when the drainage is very small with a very low discharge and it does not bring heavy silt load. An inlet-outlet structure is provided when the drainage and the canal are almost at the same level and the discharge in the drainage is small. The drainage water is admitted into the canal at a suitable site, and the excess water is discharged out the canal through an outlet provided on the canal at some distance downstream of the junction. However, the relative levels of the canal and the drainage can be altered to some extent by changing the canal alignment to have another crossing. In that case, the most suitable type of the cross-drainage work will be selected depending upon the levels at the changed crossing. An aqueduct is preferred to a siphon aqueduct. Likewise, a superpassage is preferred to a canal siphon. In the

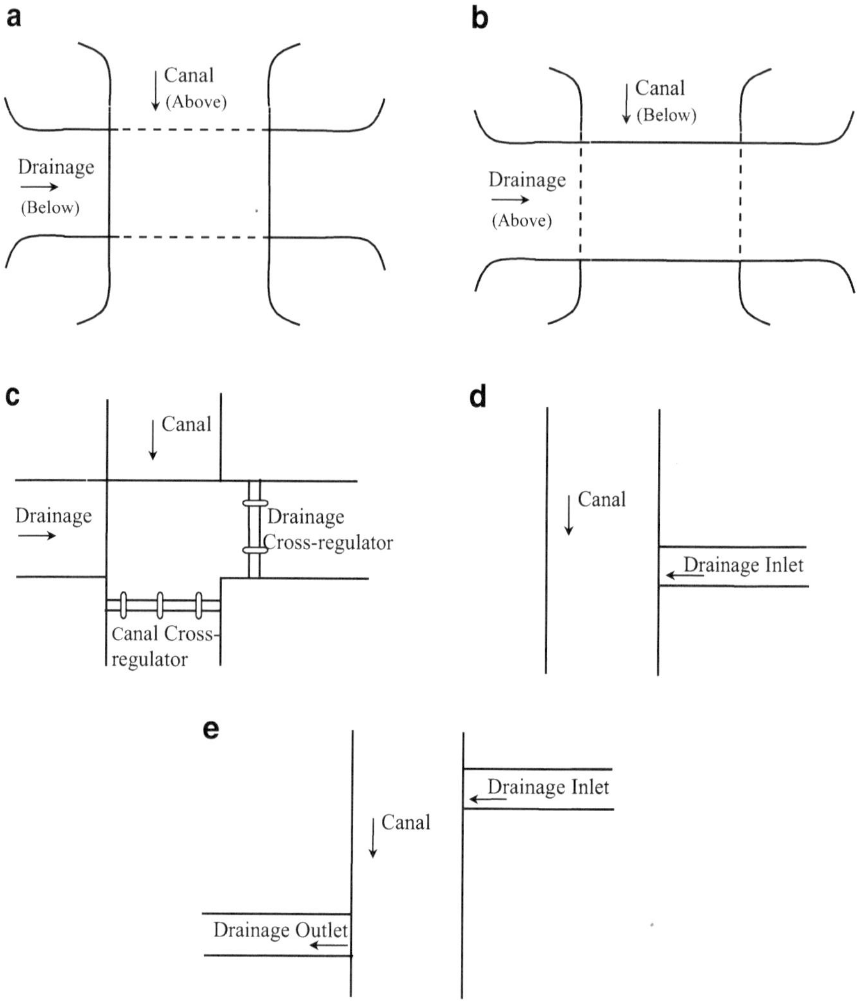

Fig. 4.5 Different types of cross-drainage structures: (**a**) aqueduct or siphon aqueduct, (**b**) superpassage or canal siphon, (**c**) level crossing, (**d**) inlet, (**e**) inlet-outlet

case of a siphon aqueduct and a canal siphon, silting problems usually occur at the crossing. Moreover, in the case of a canal siphon, there is considerable loss of command due to loss of head in the canal. The performance of inlet-outlet structures is not good.

If the natural ground slope is steeper than the required canal bed slope, then fall/drop structures are needed. When size and/or shape of the cross section of a canal changes, a transition structure is required between two sections. A transition is a structure of short length smoothly connecting two different sections.

References

Arora KR (2007) Irrigation, water power and water resources engineering. Standard Publishers Distributors, Delhi

Chahar BR (2005) Optimal design of a parabolic channel section. J Irrig Drain Eng ASCE 131(6):546–554

Chahar BR, Basu S (2009) Optimal design of curved bed Trapezoidal Canal sections. J Water Manage, Proc Inst Civil Eng, Thomas Telford 162(WM3):233–240

Chahar BR, Ahmed N, Godara R (2007) Optimal parabolic section with freeboard. J Indian Water Work Assoc 39(1):43–48

Chow VT (1973) Open channel hydraulics. McGraw Hill, New York

Hooja R (1993) The Indira Gandhi canal project in brief. J Irrig Power, Central Board of Irrigation and Power, New Delhi, India, 50(4): 9–20

Laycock A (2007) Irrigation systems: design, planning and construction. Commonwealth Agricultural Bureaux International (CABI), Oxfordshire

Loganathan GV (1991) Optimal design of parabolic canals. J Irrig Drain Eng ASCE 117(5):716–735

Sharma HD, Chawla AS (1975) Manual of canal lining, Technical report no.14, Central Board of Irrigation and Power, New Delhi

Subramanya K (2009) Flow in open channels. Tata McGraw-Hill, New Delhi

Chapter 5
Design for Minimum Flow Area

Abstract A canal based on minimization of flow area objective function for a specified discharge and canal bed slope implies maximum flow velocity or best hydraulic section. The best hydraulic section provides maximum carrying capacity for a fixed cross-sectional area or minimum cross-sectional area and perimeter to pass a given discharge. A particular set of geometric proportions yield a best hydraulic section for the specific shape of channel. As it is a minimum area and minimum perimeter section, it provides maximum hydraulic radius and requires least cost in excavation and lining. In this chapter, design of minimum flow area or maximum flow velocity sections for viscous and turbulent flows and sediment carrying canals are covered. Both the objective function and constraint for minimum flow area sections are nonlinear, so optimization problem is hard to solve analytically. The problem was converted in the unconstrained form through penalty function. A nondimensional parameter approach has been used to simplify the analysis. The dimensionless augmented function was minimized using a grid search algorithm. Using results of the optimization procedure and error minimization, close approximations in explicit form have been obtained for optimal channel dimensions. The optimization method has resulted in the optimal geometric properties for circular, trapezoidal, triangular, and rectangular sections. Presented equations result in section dimensions in single-step computations. A design example involving different cases has been presented to demonstrate the simplicity of the method.

Keywords Minimum area • Best hydraulic section • Most efficient section • Explicit design equations

Minimum flow area for a specified discharge and canal bed slope implies maximum flow velocity or best hydraulic section. Such a section also has maximum hydraulic radius or minimum wetted perimeter. In this chapter, maximum flow velocity sections for viscous and turbulent flows and sediment-carrying canals are considered.

P.K. Swamee, B.R. Chahar, *Design of Canals*, Springer Transactions in Civil and Environmental Engineering, DOI 10.1007/978-81-322-2322-1_5

5.1 Turbulent Flow Canals

The minimum area problem for a triangular canal is stated as

$$\text{Minimize} A_* = m y_n^2 \tag{5.1.1}$$

$$\text{Subject to } \phi = 1.737 \frac{m^{1.5} y_{n*}^{2.5}}{(1+m^2)^{0.25}} \ln\left[\frac{\varepsilon_*(1+m^2)^{0.5}}{6 m y_{n*}} + 0.625 \nu_* \frac{(1+m^2)^{0.75}}{(m y_{n*})^{1.5}}\right] + 1 = 0 \tag{5.1.2}$$

$$y_{n*} = y_n \left(g S_o / Q^2\right)^{0.2} \tag{5.1.3}$$

$$\varepsilon_* = \varepsilon \left(g S_o / Q^2\right)^{0.2} \tag{5.1.4}$$

$$\nu_* = \nu \left(Q^3 g S_o\right)^{-0.2} \tag{5.1.5}$$

Combining Eqs. (5.1.1) and (5.1.2), the problem is converted to the following unconstrained form:

$$\text{Minimize} \quad \Phi = A_* + p\phi^2 \tag{5.1.6}$$

where Φ = augmented function and p = penalty coefficient. Selecting ε_* and ν_* and taking p to be small and varying m and y_{n*}, the augmented function is minimized by a minimization algorithm (e.g., grid search algorithm). Starting from this minimum and increasing the p by ten times, a better minimum is found. Further, increasing the p by ten times and searching around a new minimum, the process is repeated till ϕ becomes very small in comparison to A_*. Thus, the solution of the problem is obtained.

Following this procedure, the problem was solved for a large number of values of ε_* and ν_*. The results can be expressed in the form of the following equations (Swamee 1995):

$$m^* = 1 \tag{5.1.7}$$

$$y_n^* = 0.507 L \tag{5.1.8}$$

where superscript * indicates optimality and L = a length scale given by

$$L = \left[\varepsilon \left(\frac{Q^2}{g So}\right)^{4.8} + \frac{8 \nu Q^{9.4}}{(g S_o)^{5.2}}\right]^{0.04} \tag{5.1.9}$$

It shows that L is a lumped response to design discharge, canal bed slope, roughness properties of canal surface, and fluid resistance property. Using $m^* = 1$ and $y_n^* = 0.507L$, other optimal dependent parameters are found as

$$P^* = 2y_n^* \sqrt{2} = 1.4227L \tag{5.1.10}$$

$$A^* = y_n^{*2} = 0.25302L^2 \tag{5.1.11}$$

$$V^* = Q/A^* = 3.95227QL^{-2} \tag{5.1.12}$$

The equality constraint (or resistance equation) in nondimensional form for a rectangular section is

$$2.457\frac{(b_* y_{n*})^{3/2}}{(b_* + 2y_{n*})^{1/2}} \ln\left(\frac{\varepsilon_* (b_* + 2y_{n*})}{12 b_* y_{n*}} + \frac{0.221\nu_*(b_* + 2y_{n*})^{3/2}}{(b_* y_{n*})^{3/2}}\right) + 1 = 0 \tag{5.1.13}$$

for a trapezoidal section it is

$$2.457\frac{\left(my_{n*}^2 + b_* y_{n*}\right)^{3/2}}{\left(b_* + 2y_{n*}\sqrt{1+m^2}\right)^{1/2}} \ln\left(\frac{\varepsilon_* \left(b_* + 2y_{n*}\sqrt{1+m^2}\right)}{12\left(my_{n*}^2 + b_* y_{n*}\right)} + \frac{0.221\nu_*\left(b_* + 2y_{n*}\sqrt{1+m^2}\right)^{3/2}}{\left(my_{n*}^2 + b_* y_{n*}\right)^{3/2}}\right) + 1 = 0 \tag{5.1.14}$$

and for a circular section it is

$$0.1536\frac{D_*^{2.5}(\vartheta - \sin\vartheta)^{3/2}}{(\vartheta)^{1/2}} \ln\left(\frac{\varepsilon_* \vartheta}{3D_*(\vartheta - \sin\vartheta)} + \frac{1.768\nu_*(\vartheta)^{3/2}}{D_*^{1.5}(\vartheta - \sin\vartheta)^{3/2}}\right) + 1 = 0 \tag{5.1.15}$$

where

$$b_* = b\left(gS_o/Q^2\right)^{0.2} \tag{5.1.16}$$

$$D_* = D\left(gS_o/Q^2\right)^{0.2} \tag{5.1.17}$$

The results obtained using a method similar to a triangular section can be expressed in the form of the following equations (Swamee 1995) for all the shapes:

Table 5.1 Properties of optimal channel sections for turbulent flow

Section shape	Side slope	Section shape coefficients				
	m	k_b or k_D	k_y	k_P	k_A	k_V
(1)	(2)	(3)	(4)	(5)	(6)	(7)
Triangular	1.0000	0.0000	0.5070	1.4340	0.2570	3.8904
Rectangular	0.0000	0.7170	0.3585	1.4340	0.2570	3.8904
Trapezoidal	0.5774	0.4369	0.3784	1.3108	0.2489	4.0177
Circular	—[a]	0.7850	0.3925	1.2331	0.2420	4.1322

[a]Not applicable

$$b^* = k_b L \tag{5.1.18}$$

$$D^* = k_D L \tag{5.1.19}$$

$$y_n^* = k_y L \tag{5.1.20}$$

$$P^* = k_P L \tag{5.1.21}$$

$$A^* = k_A L^2 \tag{5.1.22}$$

$$V^* = k_V Q L^{-2} \tag{5.1.23}$$

where k_y, k_P, k_A, and k_V = section shape coefficients. The value of k_V can be obtained by $k_V = 1/k_A$. The section shape coefficients, as obtained for various channel sections, are listed in Table 5.1. A perusal of Table 5.1 shows that a trapezoidal section has $m = 0.57735$. Sometimes, due to stability of side slope, the side slope cannot be kept as $m^* = 0.57735$. In such a case, the side slope m is fixed from the consideration of stability. Considering m to be constant, the trapezoidal section converts into two parameter (b and y_n) sections and the optimization is carried out. In such a case, the following equations are obtained:

$$b^* = 0.930 \frac{\sqrt{1+m^2} - m}{\left(2\sqrt{1+m^2} - m\right)^{0.375}} L \tag{5.1.24}$$

$$y_n^* = 0.465\left(2\sqrt{1+m^2} - m\right)^{-0.375} L \tag{5.1.25}$$

$$P^* = 0.930\left(2\sqrt{1+m^2} - m\right)^{0.625} L \tag{5.1.26}$$

$$A^* = 0.216\left(2\sqrt{1+m^2} - m\right)^{0.25} L^2 \qquad (5.1.27)$$

Thus, if it is not desirable to construct the canal on the side slope at optimal value due to any practical reason, then the optimal section parameters are functions of m.

Example 5.1 Design a trapezoidal canal section having $m = 1$ for carrying a discharge of 25 m³/s on a longitudinal slope of 0.0002. The canal lining has $\varepsilon = 1$ mm. The water temperature is 20 °C at which kinematic viscosity of water is 1.1×10^{-6} m²/s.

Adopting $g = 9.8$ m/s² and using Eq. (5.1.9), $L = 8.66$ m. Further, using Eqs. (5.1.18) and (5.1.20), $b^* = 2.67$ m and $y_n^* = 3.20$ m. The flow area $A^* = y_n^*\left(b^* + y_n^*\right) = 18.874$ m² yielding $V^* = 25/18.874 = 1.33$ m/s.

5.2 Viscous Flow Channels

Using Eqs. (2.1.2) and (3.1.5), the design of a rectangular section for viscous flow is formulated as

$$\text{Minimize} \quad A = by_n \qquad (5.2.1)$$

Subject to
$$\phi = \frac{gS_o b y_n^3}{3\nu Q}\left[1 - \frac{384}{\pi^5}\frac{y_n}{b}\sum_{i=0}^{\infty}(2i+1)^{-5}\tanh\left(\frac{2i+1}{4}\frac{\pi b}{y_n}\right)\right] - 1 = 0 \qquad (5.2.2)$$

These two variable minimization problems have been solved using the penalty function method. The resulting optimal dimensions are expressed in the following forms (Swamee 2000):

$$b^* = 2.746L \qquad (5.2.3)$$

$$y_n^* = 1.373L \qquad (5.2.4)$$

where $L =$ a length scale given by

$$L = \left(\frac{\nu Q}{gS_o}\right)^{0.25} \qquad (5.2.5)$$

An analysis of other optimal sections revealed the following functional forms similar to turbulent flow channels:

Table 5.2 Properties of optimal channel sections for viscous flow

Section shape	Side slope	Section shape coefficients				
	m	k_b or k_D	k_y	k_P	k_A	k_V
(1)	(2)	(3)	(4)	(5)	(6)	(7)
Triangular	1.000	0.000	1.942	5.492	3.771	0.265
Rectangular	0.000	2.746	1.373	5.492	3.771	0.265
Trapezoidal	0.577	1.673	1.449	5.020	3.638	0.275
Circular	—[a]	3.004	1.502	4.720	3.535	0.282

[a]Not applicable

$$b^* = k_b L \tag{5.2.6}$$

$$D^* = k_D L \tag{5.2.7}$$

$$y_n^* = k_y L \tag{5.2.8}$$

$$P^* = k_P L \tag{5.2.9}$$

$$A^* = k_A L^2 \tag{5.2.10}$$

$$V^* = k_V Q L^{-2} \tag{5.2.11}$$

The section shape coefficients, as obtained for various channel sections, are listed in Table 5.2. In this case also the trapezoidal section has $m = 0.5773$. If predetermined side slope due to practical considerations is adopted, then similar to turbulent flow case, the following equations are obtained:

$$b^* = 3.561 \frac{\sqrt{1+m^2} - m}{\left(2\sqrt{1+m^2} - m\right)^{0.375}} L \tag{5.2.12}$$

$$y_n^* = 1.781\left(2\sqrt{1+m^2} - m\right)^{-0.375} L \tag{5.2.13}$$

$$P^* = 3.561\left(2\sqrt{1+m^2} - m\right)^{0.625} L \tag{5.2.14}$$

$$A^* = 3.171\left(2\sqrt{1+m^2} - m\right)^{0.25} L^2 \tag{5.2.15}$$

Example 5.2 The objective is to design a rectangular channel section for carrying a discharge of 0.025 m^3/s on a slope of 0.005. The kinematic viscosity of fluid is

4×10^{-3} m^2/s. Adopting g = 9.8 m/s^2 and using Eq. (5.2.5), $L = 0.213$ m. Further, using Eqs. (5.2.3) and (5.2.4), $b^* = 0.585$ m and $y_n^* = 0.293$ m. The flow area $A^* = b^* y_n^* = 0.1714$ m^2. Thus, $V^* = 0.025/0.1714 = 0.146$ m/s.

5.3 Sediment-Transporting Channels

In the present investigation, the flow area is minimized. Thus, the sections obtained are maximum sediment velocity sections. As per Eq. (3.1.25) V_s varies directly with the hydraulic radius, the design problem reduces to the maximization of R.

The canal should be able to transport the requisite sediment discharge. Thus, the design is subjected to the constraint of Eq. (3.1.22) written as

$$A\sqrt{R} = K \tag{5.3.1}$$

where

$$K = \frac{81(s-1)^{1.5} f^2 Q_s}{2C_D^{0.75} S_o^2 \sqrt{g}} \tag{5.3.2}$$

For a trapezoidal channel section (see Fig. 2.1c), the hydraulic radius is written as

$$R = \frac{b\alpha\,(1+\alpha)}{m + 2\alpha\sqrt{1+m^2}} \tag{5.3.3}$$

where $\alpha = my_n/b$. On the other hand, Eq. (5.3.1) is written as

$$K = \frac{b^2}{m}\alpha\,(1+\alpha)\,R^{1/2} \tag{5.3.4}$$

Eliminating α between Eqs. (5.3.3) and (5.3.4), the following equation is obtained:

$$(\cos\theta - 2)\,b^2R^3 - 4KR^{5/2}\text{cosec}\theta + 2\,(1-\cos\theta)\,KbR^{3/2} + K^2\cos\theta = 0 \tag{5.3.5}$$

where $\theta = \cot^{-1}m$. For maximum R differentiating Eq. (5.3.5) with respect to b and θ and putting $\partial R/\partial b = \partial R/\partial\theta = 0$ and simplifying, the following equations are obtained:

$$b = \frac{1-\cos\theta}{2-\cos\theta}\frac{K}{R^{3/2}} \tag{5.3.6}$$

$$b^2R^3 - 4KR^{5/2}\text{cosec}^2\theta\;\cot\theta - 2\,KbR^{3/2} + K^2 = 0 \tag{5.3.7}$$

Eliminating b between Eqs. (5.3.6) and (5.3.7)

$$K = 4 \cot\theta \ (2 \text{ coesc}\theta - \cot\theta)^2 R^{5/2} \tag{5.3.8}$$

Eliminating b and R between equations (5.3.5), (5.3.6), and (5.3.8), $\cos\theta = 1/2$ giving $\theta^* = \pi/3$ or $m* = 1/\sqrt{3}$. Substituting $\theta^* = \pi/3$ in Eq. (5.3.8), the maximum hydraulic radius R^* is

$$R^* = \left(\frac{K}{4\sqrt{3}}\right)^{0.4} \tag{5.3.9}$$

Putting $\theta = \pi/3$ and using Eq. (5.3.9), Eq. (5.3.6) gives

$$b^* = \left(\frac{8K}{3^{7/4}}\right)^{0.4} \tag{5.3.10}$$

Putting $m^* = 1/\sqrt{3}$, and using Eqs. (5.3.9) and (5.3.10), Eq. (5.3.4) yields $\alpha^* = 1/2$. Thus, using Eq. (5.3.10) and the definition of α, one gets

$$y^* = \left(\frac{\sqrt{2}K}{\sqrt{3}}\right)^{0.4} \tag{5.3.11}$$

Knowing m^*, b^*, and $y,^*$ the optimal area is obtained as

$$A^* = \left(2^{0.5} 3^{0.125} K\right)^{0.4} \tag{5.3.12}$$

Using Eqs. (5.3.9) and (5.3.12), the optimal perimeter P^* is found to be

$$P^* = \left(2^3 3^{0.75} K\right)^{0.4} \tag{5.3.13}$$

For other optimal open-channel sections, the functional form is the same as given by Eqs. (5.1.18), (5.1.19), (5.1.20), (5.1.21), (5.1.22) and (5.1.23) where L is given by

$$L = \left[\frac{(s-1)^{1.5} f^2 Q_s}{C_D^{0.75} S_o^2 \sqrt{g}}\right]^{0.4} \tag{5.3.14}$$

The section shape coefficients for various channel shapes are given in Table 5.3.

The optimal trapezoidal channel has side slope $m = 0.5773$. If for any practical reason it is not desirable to construct the canal on $m = 0.5773$, then the section shape coefficients will be functions of m. For such a case, the following equations are obtained:

Table 5.3 Properties of sediment-transporting optimal channel sections

Section shape	Side slope	Section shape coefficients					
	m	k_b or k_D	k_y	k_P	k_R	k_A	k_{Vs}
	(2)	(3)	(4)	(5)	(6)	(7)	(8)
Triangular	1.0000	0.0000	5.4112	15.3051	1.9131	29.2807	0.0342
Rectangular	0.0000	7.6525	3.8263	15.3051	1.9131	29.2807	0.0342
Trapezoidal	0.5774	4.6799	4.0529	14.0396	2.0264	28.4503	0.0351
Circular	—[a]	8.4289	4.2144	13.2401	2.1072	27.8997	0.0358

[a]Not applicable

$$k_b = 10.0976\left(\sqrt{1+m^2}-m\right)\left(2\sqrt{1+m^2}-m\right)^{-0.4} \tag{5.3.15}$$

$$k_y = 5.0488\left(2\sqrt{1+m^2}-m\right)^{-0.4} \tag{5.3.16}$$

$$k_P = 10.0976\left(2\sqrt{1+m^2}-m\right)^{0.6} \tag{5.3.17}$$

$$k_R = 2.5244\left(2\sqrt{1+m^2}-m\right)^{-0.4} \tag{5.3.18}$$

$$k_A = 25.4904\left(2\sqrt{1+m^2}-m\right)^{0.2} \tag{5.3.19}$$

The value of k_V can be obtained by $k_V = 1/k_A$.

Example 5.3 The objective is to design a rectangular channel section for carrying a sediment discharge of 0.01 m^3/s with $s = 2.65$ and $d = 0.1$ mm on a slope of 0.01.

In the design $g = 9.8$ m/s^2, $\varepsilon = 1$ mm and $\nu = 1 \times 10^{-6}$ m^2/s have been adopted. Thus, $\nu_* = \nu/\left[d\sqrt{(s-1)gd}\right] = 0.2487$. Following Swamee and Ojha (1991), the fall velocity w is

$$w = \sqrt{(s-1)gd}\left\{\left[(18\nu_*)^2 + (18\nu_*)^{0.54}\right]^5 + \left[(10^8\nu_*)^{1.7} + 1.43\times10^6\right]^{-0.346}\right\}^{-0.1} \tag{5.3.20}$$

Equation (5.3.20) gives $w = 0.008077$ m/s. The particle Reynolds number $\mathbf{R}_s = wd/\nu = 0.8077$. Further, following Swamee and Ojha (1991), the drag coefficient is

Table 5.4 Design iterations

Iteration number	f	b (m)	y_n (m)	Q (m^3/s)	V (m/s)	$\mathbf{R}$
(1)	(2)	(3)	(4)	(5)	(6)	(7)
1.	0.0100	0.3640	0.1821	0.1153	1.8892	0.6880×10^6
2.	0.0273	0.8137	0.4069	0.5562	1.7085	0.1391×10^7
3.	0.0220	0.6843	0.3422	0.3989	1.7460	0.1195×10^7
4.	0.0230	0.7093	0.3546	0.4273	1.7381	0.1233×10^7

$$C_D = 0.5\left\{16\left[\left(\frac{24}{\mathbf{R_s}}\right)^{1.6} + \left(\frac{130}{\mathbf{R_s}}\right)^{0.72}\right]^{2.5} + \left[\left(\frac{40,000}{\mathbf{R_s}}\right)^{2} + 1\right]^{-0.25}\right\}^{0.25} \tag{5.3.21}$$

Using Eq. (5.3.21), $C_D = 32.7890$. For starting the algorithm, f is assumed to be 0.01. For a rectangular section adopting $m = 0$ and using Eqs. (3.1.18), (3.1.23), (3.1.26), (3.1.27), (5.1.18), (5.1.20), (5.3.14), (5.3.15), and (5.3.16), iterations are carried out. These iterations are shown in Table 5.4.

Thus, a bed width of 0.75 m is provided. Allowing adequate freeboard canal depth of 0.5 m is provided.

Exercises

1. Design a rectangular canal of bed slope 0.005 carrying a discharge of 25 m^3/s.
2. Design a trapezoidal canal of bed slope 0.001 having a discharge of 225 m^3/s.
3. Design a rectangular canal of bed slope 0.01 carrying a discharge of 0.025 m^3/s of glycerin.

References

Swamee PK (1995) Optimal irrigation canal sections. J Irrig Drain Eng ASCE 121(6):467–469

Swamee PK (2000) Optimal open channel sections for viscous flow. ISH J Hydraul Eng 6(2):40–44

Swamee PK, Ojha CSP (1991) Drag coefficient and fall velocity of non-spherical particles. J Hydraul Eng ASCE 117(5):660–667

Chapter 6
Minimum Cost Canal Section

Abstract Though the minimum area section is generally adopted for lined canals, it is not the minimum cost section as it does not involve lining cost and the cost of earthwork. In general, the cost of earthwork varies with canal depth. Design of a minimum cost canal section involves minimization of the sum of earthwork cost and cost of lining subject to uniform flow condition in the canal, which results in nonlinear objective function and nonlinear equality constraint making the problem hard to solve analytically. This chapter discusses the nonlinear optimization method to obtain explicit design equations and section shape coefficients for the design variables for minimum cost canal section for triangular, rectangular, trapezoidal, and circular shapes. These optimal design equations and coefficients have been obtained by analyzing a very large number of optimal sections resulted from the application of optimization procedure in the wide application ranges of input variables. The analysis consists of conceiving an appropriate functional form and then minimizing errors between the optimal values and the computed values from the conceived function with coefficients. Particular cases like minimum earthwork cost section and minimum and maximum discharge canal sections are also included in the chapter. The optimal design equations show that on account of additional cost of excavation with canal depth, the optimal section is wider and shallower than the minimum area section. Application of the proposed design equations along with the tabulated section shape coefficients results directly into the optimal dimensions of minimum cost canal sections without going through the conventional trial and error method of canal design.

Keywords Minimum cost canal section • Minimum earthwork cost section • Minimum cost lined section • Minimum and maximum discharge section • Section shape coefficient • Explicit design equations

6.1 Construction Cost Minimization

As indicated by Eqs. (2.3.1) and (2.4.2), the section cost consists of the lining cost and the earthwork cost. Adding Eqs. (2.3.1) and (2.4.2), the section cost C is written as

P.K. Swamee, B.R. Chahar, *Design of Canals*, Springer Transactions in Civil and Environmental Engineering, DOI 10.1007/978-81-322-2322-1_6

$$C = c_L P + c_e A + c_r A\overline{y} \tag{6.1.1}$$

The section cost as given by Eq. (6.1.1) has to be minimized subject to the constraint described by Eq. (3.1.15). For a triangular channel, Eq. (6.1.1) is written as

$$\text{minimize } C_* = 2c_{L*}y_{n*}\sqrt{1+m^2} + my_n^2 + \frac{c_{r*}my_{n*}^3}{3} \tag{6.1.2}$$

subject to constraint given by Eq. (5.1.2)
where

$$C_* = \frac{C}{c_e}\left(\frac{gS_o}{Q^2}\right)^{0.4} \tag{6.1.3}$$

$$c_{L*} = \frac{c_L}{c_e}\left(\frac{gS_o}{Q^2}\right)^{0.2} \tag{6.1.4}$$

$$c_{r*} = \frac{c_r}{c_e}\left(\frac{Q^2}{gS_o}\right)^{0.2} \tag{6.1.5}$$

Using the exterior penalty function method as described in Sect. 5.1 and varying $10^{-6} \le \varepsilon_* \le 10^{-3}$, $10^{-7} \le \nu_* \le 10^{-5}$, $0 \le c_{L*} \le \infty$, and $0 \le c_{r*} \le 50c_{L*}$, a large number of optimum sections were obtained. An analysis of these sections gave the following relationships:

$$m^* = \frac{15.0491c_L + c_eL + 0.3039c_rL^2}{15.0491c_L + c_eL} \tag{6.1.6}$$

$$y_n^* = 0.5070\frac{15.0389c_L + c_eL}{15.0389c_L + c_eL + 0.13973c_rL^2}L \tag{6.1.7}$$

$$C^* = 1.4249c_LL + 0.2530c_eL^2 + 0.0397c_rL^3 \tag{6.1.8}$$

where L is the length scale defined by Eq. (5.1.9). Following the above procedure for rectangular, trapezoidal, and circular canal sections, generalized optimal equations for all the four canal shapes are expressed as (Swamee et al. 2000)

$$m^* = k_{m0}\frac{k_{mL}c_L + c_eL + k_{mr}c_rL^2}{k_{mL}c_L + c_eL} \tag{6.1.9}$$

$$b^* = k_{b0}\frac{k_{bL}c_L + c_eL + k_{br}c_rL^2}{k_{bL}c_L + c_eL}L \tag{6.1.10}$$

$$D^* = k_{D0}\frac{k_{DL}c_L + c_eL + k_{Dr}c_rL^2}{k_{DL}c_L + c_eL}L \tag{6.1.11}$$

Table 6.1 Properties of optimal canal sections

Entity	Section shape coefficients	Section shapes			
		Triangular	Rectangular	Trapezoidal	Circular
(1)	(2)	(3)	(4)	(5)	(6)
Side slope	k_{m0}	1.0000		0.5774	
	k_{mL}	15.0491		14.2772	
	k_{mr}	0.3039		0.2163	
Bed width or diameter	k_{b0} or k_{D0}		0.7114	0.4341	0.7807
	k_{bL} or k_{DL}		15.0284	14.2425	13.6232
	k_{br} or k_{Dr}		0.3201	0.3484	0.2482
Normal depth	k_{y0}	0.5030	0.3557	0.3759	0.3903
	k_{yL}	15.0389	15.0234	14.2274	12.9379
	k_{yr}	0.1397	0.3066	0.2233	0.1263
Cost	k_{ce}	0.2530	0.2530	0.2448	0.2393
	k_{cL}	1.4249	1.4249	1.3037	1.2265
	k_{cr}	0.0397	0.0396	0.0372	0.0371

$$y_n^* = k_{y0}\frac{k_{yL}c_L + c_e L}{k_{yL}c_L + c_e L + k_{yr}c_r L^2}L \tag{6.1.12}$$

$$C^* = k_{cL}c_L L + k_{ce}c_e L^2 + k_{cr}c_r L^3 \tag{6.1.13}$$

Table 6.1 lists section shape coefficients.

6.1.1 Minimum and Maximum Discharge Sections

Equation (5.1.9) indicates that for $Q \to 0$, $L \to 0$. Thus, considering $L \to 0$ for small discharges, Eqs. (6.1.9), (6.1.10), (6.1.11) and (6.1.12) reduce to

$$m^* = k_{m0} \tag{6.1.14}$$

$$b^* = k_{b0}L \tag{6.1.15}$$

$$D^* = k_{D0}L \tag{6.1.16}$$

$$y_n^* = k_{y0}L \tag{6.1.17}$$

That is, for small discharges, the optimal section reduces to minimum area section. On the other hand, for very large discharge, $Q \to \infty$, $L \to \infty$. Thus, putting $L \to \infty$ for very large discharge, Eqs. (6.1.9), (6.1.10) and (6.1.12) reduce to

$$m^* \to \infty \tag{6.1.18}$$

$$b^* \rightarrow \left(k_{b0} + \frac{k_{br} c_r}{c_e} \right) L \tag{6.1.19}$$

$$y_n^* \rightarrow \frac{k_{y0} c_e}{k_{yr} c_r} \tag{6.1.20}$$

A perusal of Eqs. (6.1.18) and (6.1.19) reveal that for a large discharge, the canal section becomes flatter. On the other hand, Eq. (6.1.20) indicates that with the increase in discharge, the normal depth asymptotically approaches to a constant value that depends on section shape coefficients and cost coefficients. For ordinary soil, Table 2.2 gives $c_e/c_r = 7$ m. Further, Table 6.1 for rectangular section gives $k_{y0}/k_{yr} = 0.3557/0.3066 = 1.160$. With these data, Eq. (6.1.20) gives $y_n^* = 1.160 \times 7 = 8.12$ m, which is the maximum normal depth.

6.1.2 Minimum Earthwork Cost Section

A canal passing through hard strata may be kept unlined; however, it is designed as a rigid boundary canal. Dropping the lining cost coefficient c_L in Eqs. (6.1.9), (6.1.10), (6.1.11), (6.1.12) and (6.1.13), the following equations are obtained:

$$m^* = k_{m0} \frac{c_e + k_{mr} c_r L}{c_e} \tag{6.1.21}$$

$$b^* = k_{b0} \frac{c_e + k_{br} c_r L}{c_e} L \tag{6.1.22}$$

$$D^* = k_{D0} \frac{c_e + k_{Dr} c_r L}{c_e} L \tag{6.1.23}$$

$$y_n^* = \frac{k_{y0} c_e L}{c_e + k_{yr} c_r L} \tag{6.1.24}$$

$$C^* = k_{ce} c_e L^2 + k_{cr} c_r L^3 \tag{6.1.25}$$

Equations (6.1.9), (6.1.10), (6.1.11), (6.1.12), (6.1.21), (6.1.22), (6.1.23), and (6.1.24) indicate that for $c_r = 0$, the optimal section is the minimum area section. However, with the increase in c_r, the canal section becomes wide and shallow. On the other hand, with the increase in c_e and/or c_L, the canal section approaches to the corresponding minimum area section. Similarly, the case $c_r = 0$ also represents minimum wetted perimeter section. Thus, the minimum area section and the minimum wetted perimeter section are the same.

Example 6.1 Design a concrete-lined trapezoidal canal section for carrying a discharge of 125 m^3/s on a longitudinal slope of 0.0002. The canal passes through a stratum of ordinary soil for which $c_e/c_r = 7$ m. Further, it is proposed to provide concrete lining with $c_L/c_e = 12$ m.

Solution For the design, $g = 9.80$ m/s^2, $\nu = 1 \times 10^{-6}$ m^2/s (water at 20 °C), and $\varepsilon = 1$ mm are adopted.

Using Eq. (5.1.9), $L = 16.06$ m. For a trapezoidal channel section, Table 6.1 gives $k_{m0} = 0.5774$, $k_{mL} = 14.2772$, $k_{mr} = 0.2163$, $k_{b0} = 0.4341$, $k_{bL} = 14.2425$, $k_{br} = 0.3484$, $k_{y0} = 0.3759$, $k_{yL} = 14.2274$, $k_{yr} = 0.2233$, $k_{c0} = 0.2448$, $k_{cL} = 1.3037$, and $k_{cr} = 0.0372$.

With these coefficients, Eqs. (6.1.9), (6.1.10), (6.1.12) and (6.1.13) yielded

$$\begin{aligned} m^* &= k_{m0}\frac{k_{mL}(c_L/c_e) + L + k_{mr}(c_r/c_e)L^2}{k_{mL}(c_L/c_e) + L} \\ &= 0.5774\frac{14.2772 \times 12 + 16.06 + 0.2163 \times (1/7) \times 16.06^2}{14.2772 \times 12 + 16.06} \\ &= 0.602. \end{aligned}$$

$$\begin{aligned} b^* &= k_{b0}\frac{k_{bL}(c_L/c_e) + L + k_{br}(c_r/c_e)L^2}{k_{bL}(c_L/c_e) + L}L \\ &= 0.4341\frac{14.2425 \times 12 + 16.06 + 0.3484 \times (1/7) \times 16.06^2}{14.2425 \times 12 + 16.06}16.06 = 7.45\text{ m.} \\ y_n^* &= k_{y0}\frac{k_{yL}(c_L/c_e) + L}{k_{yL}(c_L/c_e) + L + k_{yr}(c_r/c_e)L^2}L \\ &= 0.3759\frac{14.2274 \times 12 + 16.06}{14.2274 \times 12 + 16.06 + 0.2233 \times (1/7) \times 16.06^2}16.06 = 5.782\text{ m.} \\ C^* &= k_{cL}(c_L/c_e)Lc_e + k_{ce}L^2c_e + k_{cr}(c_r/c_e)L^3c_e \\ &= 1.3037 \times 12 \times 16.06c_e + 0.2448 \times 16.06^2c_e + 0.0372 \times (1/7) \times 16.06^3c_e \\ &= 251.25c_e + 63.14c_e + 22.01c_e = 251.25c_e + 85.15c_e = 336.40c_e \end{aligned}$$

It can be seen that lining cost per meter length $C_L = 251.25c_e$; and the excavation cost per meter length $C_e = 63.14c_e + 22.01c_e = 85.160c_e$. Thus, it can be seen that lining shares a major portion of total cost. Adopting $m = 0.6$, $b = 7.45$ m, and $y_n{}^* = 5.80$ m yields $A = y_n(b + my_n) = 5.8(7.45 + 0.6 \times 5.8) = 63.394$ m^2. Thus, $V = 125/63.394 = 1.972$ m/s, which is within permissible limit (Table 4.1).

6.2 Generalized Equations of Wider Applicability

Equations (6.1.21), (6.1.22), (6.1.23), (6.1.24), and (6.1.25) are applicable for $c_rL/c_e \leq 1$. For higher values of additional earthwork cost, i.e., $c_rL/c_e > 2$, the optimal design parameters are nonlinear functions of length scale. In the range $0 \leq c_rL/c_e \leq 100$, the optimal dimensions and corresponding cost of a least earthwork cost section become (Swamee et al. 2001)

Table 6.2 Coefficients and exponents for minimum earthwork cost sections

Entity	Coefficients or exponents	Section shapes			
		Triangular	Rectangular	Trapezoidal	Circular
(1)	(2)	(3)	(4)	(5)	(6)
Side slope	t_{mr}	0.2572		10.000	
	r_{mr}	0.8613		1.2586	
	s_{mr}	1.1525		0.08069	
Bed width or diameter	t_{br} or t_{Dr}		0.1122	0.3195	0.07023
	r_{br} or r_{Dr}		0.7143	0.9342	0.9014
	s_{br} or s_{Dr}		2.3759	1.0712	3.2330
Normal depth	t_{yr}	0.02797	0.2550	0.2088	0.0161
	r_{yr}	0.7127	0.7422	0.8123	0.7196
	s_{yr}	4.3600	1.0375	1.0280	6.0000
Cost	t_{cr}	0.2451	0.2234	0.2277	0.2124
	r_{cr}	1.0637	0.9233	0.9544	0.9744
	s_{cr}	0.6212	0.7107	0.6855	0.7440

$$m^* = k_{m0}[1 + t_{mr}(c_r L/c_e)^{r_{mr}}]^{s_{mr}} \tag{6.2.1}$$

$$b^* = k_{b0}[1 + t_{br}(c_r L/c_e)^{r_{br}}]^{s_{br}} L \tag{6.2.2}$$

$$D^* = k_{D0}[1 + t_{Dr}(c_r L/c_e)^{r_{Dr}}]^{s_{Dr}} L \tag{6.2.3}$$

$$y_n^* = k_{y0}\left[1 + t_{yr}(c_r L/c_e)^{r_{yr}}\right]^{-s_{yr}} L \tag{6.2.4}$$

$$C_e^* = k_{ce}[1 + t_{cr}(c_r L/c_e)^{r_{cr}}]^{s_{cr}} c_e L^2 \tag{6.2.5}$$

Section shape coefficients (k_{fs} and t_{fs}) and section shape exponents (r_{fs} and s_{fs}) are listed in Table 6.2.

Example 6.2 Design rectangular sections for $Q = 50$ m^3/s, $S_o = 0.0002$, $\varepsilon = 4$ mm (rough surface), $\nu = 1.007 \times 10^{-6}$ m^2/s (water at 20 °C), and $c_e/c_r = 9$ m.

Solution For the given data, $L = 11.930$ m and $c_r L/c_e = 1.326$. The section shape coefficients and exponents for a rectangular section from Table 6.2 are for bed width $k_{b0} = 0.7114$, $t_{br} = 0.1122$, $r_{br} = 0.7143$, and $s_{br} = 2.3759$; for normal depth $k_{y0} = 0.3557$, $t_{yr} = 0.2550$, $r_{yr} = 0.7422$, and $s_{yr} = 1.0375$; and for cost $k_{ce} = 0.2530$, $t_{cr} = 0.2234$, $r_{cr} = 0.9233$, and $s_{cr} = 0.7107$. Using these in design Eqs. (6.2.2), (6.2.4) and (6.2.5), $b^* = 11.519$ m, $y_n{}^* = 3.196$ m, and $C_e{}^* = 43.15c_e$. Thus, $A = 36.815$ m^2 and $V = 1.358$ m/s, which is safe.

Using Eqs. (6.1.22), (6.1.24) and (6.1.25) along with tabulated values of section shape coefficients from Table 6.1, the corresponding design values are $b^* = 12.09$ m,

$y_n^* = 3.017$ m, $C_e^* = 43.50c_e$, and $A = 36.473$ m^2. Actual optimization resulted to $b^* = 11.643$ m, $y_n^* = 3.166$ m, $C_e^* = 43.338c_e$, and $A = 36.856$ m^2. Thus, for $c_rL/c_e = 1.326 > 1$, Eqs. (6.2.2), (6.2.4) and (6.2.5) yielded better design parameters than Eqs. (6.1.22), (6.1.24) and (6.1.25).

For $c_r = 0$ (minimum area section), the design values are $b^* = 8.487$ m, $y_n^* = 4.243$ m, $C_e^* = 36.01c_e$, $A = 36.01$ m^2, and $V = 1.389$ m/s. The design reveals that there is 35.73 % increase in the bed width, 24.68 % decrease in the normal depth, and 3.03 % decrease in the canal cost when the earthwork cost as a function of depth is considered. It can be seen that the canal dimensions are more sensitive than the cost canal section to change in earthwork cost with depth of excavation.

For a triangular canal section, the optimal design parameters are $m^* = 1.387$, $y_n^* = 5.183$ m, $C_e^* = 43.01c_e$, $A = 37.260$ m^2, and $V = 1.342$ m/s. Comparison of costs between optimal triangular and rectangular canals shows that the former canal is marginal economical than the latter canal, while in minimum area section, both are the same.

Exercises

1. Design an unlined trapezoidal canal section for carrying a discharge of 250 m^3/s on a longitudinal slope of 0.0001. The canal passes through a hard stratum for which $c_e/c_r = 9$ m. Adopt $\varepsilon = 4$ mm (rough surface).
2. Design the following lined canal sections: (a) rectangular, (b) triangular, and (c) circular canal sections for $Q = 5$ m^3/s, $S_o = 0.0002$, $c_L/c_e = 12$ m, and $c_e/c_r = 9$ m. Take $\varepsilon = 2$ mm.
3. Design canal sections for $Q = 75$ m^3/s and $S_o = 0.0002$, if (a) canal is lined having $c_L/c_e = 15$ m, $\varepsilon = 1.5$ mm, and $c_e/c_r = 10$ m and (b) canal is unlined having $c_e/c_r = 6$ m and $\varepsilon = 3.5$ mm.

References

Swamee PK, Mishra GC, Chahar BR (2000) Minimum cost design of lined canal sections. Water Resour Manag Int J 14(1):1–12

Swamee PK, Mishra GC, Chahar BR (2001) Design of minimum earthwork cost canal sections. Water Resour Manag Int J 15(1):17–30

Chapter 7
Minimum Water Loss Canal Section

Abstract An irrigation canal may be an unlined canal or a lined canal. The loss of water due to seepage and evaporation from canals constitutes a substantial part of the available water. The seepage loss results not only in depleted freshwater resources but also causes water logging, salinization, groundwater contamination, and health hazards. To minimize seepage and to transport water efficiently, lined canals were envisaged. A perfect lining would prevent all the seepage loss, but canal lining deteriorates with time. The thickness of the lining material is small and cracks may develop anywhere on the perimeter. The seepage from a canal with cracked lining is likely to approach the quantity of seepage from an unlined canal. The quantity of seepage is affected by the presence of a drainage layer. This chapter deals with the design of canal sections considering seepage and evaporation water losses for triangular, rectangular, trapezoidal, parabolic, and power law canals. The chapter also includes special cases, for example, minimum seepage loss sections without drainage layer and minimum seepage loss sections with drainage layer at shallow depth. The resultant explicit equations for the design variables of minimum water loss sections have been obtained using nonlinear optimization technique. The proposed equations along with tabulated section shape parameters facilitate easy design of the minimum water loss section and computation of water loss from the section without going through the conventional and cumbersome trial and error method. Design examples have been included to demonstrate the simplicity of the method.

Keywords Water loss • Seepage loss • Drainage layer • Evaporation loss • Minimum water loss canal design • Minimum seepage loss canal section design • Optimal sections • Section shape coefficient • Explicit design equations

The loss of water due to seepage and evaporation from irrigation canals constitutes a substantial part of the usable water. By the time the water reaches the field, more than half of the water supplied at the head of the canal is lost in seepage and evaporation (Sharma and Chawla 1975). Seepage loss is the major and the most important part of the total water loss. The other part, i.e., evaporation loss, is important particularly in water-scarce areas. Considerable part of flow may be lost from a network of canals by the way of evaporation in high evaporating conditions.

P.K. Swamee, B.R. Chahar, *Design of Canals*, Springer Transactions in Civil and Environmental Engineering, DOI 10.1007/978-81-322-2322-1_7

This needs special consideration for a long channel carrying small discharge in arid regions. Thus, care must be taken in the design of such canals to account for evaporative losses along with seepage loss.

7.1 Minimum Seepage Loss Canal Sections

Using Eqs. (2.4.1) and (3.1.15), the design of minimum seepage loss canal section boils down to

$$\text{Minimize} \quad q_{s*} = y_{n*} F_s \tag{7.1.1}$$

$$\text{Subject to} \quad \Phi_* = 2.457 A_* \sqrt{R_*} \ln\left(\frac{\varepsilon_*}{12R_*} + \frac{0.221\nu_*}{R_*^{1.5}}\right) + 1 = 0 \tag{7.1.2}$$

$$\text{where} \quad q_{s*} = \frac{q_s}{k}\left(\frac{gS_o}{Q^2}\right)^{0..2} \tag{7.1.3}$$

7.1.1 Channels Having Drainage Layer at Large Depth

For a triangular section, the minimum seepage section problem as described by Eqs. (7.1.1) and (7.1.2) changes to

$$\text{Minimize} \quad F_{s*} = y_{n*}\left\{[\pi(4-\pi)]^{1.3} + (2m)^{1.3}\right\}^{0.77} \tag{7.1.4}$$

$$\text{Subject to } \Phi_* = 1.737 \frac{m^{1.5} y_{n*}^{2.5}}{(1+m^2)^{0.25}} \ln\left[\frac{\varepsilon_*(1+m^2)^{0.5}}{6my_{n*}} + 0.625\nu_* \frac{(1+m^2)^{0.75}}{(my_{n*})^{1.5}}\right] + 1 = 0 \tag{7.1.5}$$

Using an optimization method for the solution of the problem for a number of values of ε_* and ν_* in the practical range, a large number of optimal sections are generated. Making use of these optimal sections and adopting a procedure described in Sect. 5.2, the following empirical equations were obtained:

$$m^* = 1.244 \tag{7.1.6}$$

$$y_{n*} = 0.452L \tag{7.1.7}$$

Table 7.1 Properties of optimal canal sections for minimum seepage loss

Section shape	Side slope	Section shape coefficients				
	m	k_b, k_D	k_y	k_A	k_V	k_q
(1)	(2)	(3)	(4)	(5)	(6)	(7)
Triangular	1.2445	0.000	0.4518	0.2540	3.9365	2.0015
Rectangular	0.000	0.7986	0.3178	0.2538	3.9402	2.0399
Trapezoidal	0.5984	0.5446	0.3309	0.2457	4.0695	1.9227
Circular	—	1.540	0.300	0.256	3.906	1.555

$$q_s^* = 2.001kL \tag{7.1.8}$$

Following the above procedure for rectangular and trapezoidal canals, section shape coefficients [as defined by Eqs. (5.1.18), (5.1.19), (5.1.20), (5.1.21), (5.1.22) and (5.1.23)] were obtained. These section shape coefficients are listed in Table 7.1. See Swamee et al. (2000) and Swamee and Kashyap (2001). In Table 7.1, the coefficient k_q is defined by

$$q_s^* = k_q kL \tag{7.1.9}$$

A comparison of Table 5.1 with Table 7.1 reveals that minimum seepage loss canals are wider and shallower than the corresponding minimum area canals.

A perusal of Table 7.1 reveals that the optimal dimensions are independent of k, i.e., hydraulic conductivity of the seepage layer through which a canal passes. This is so because for 2D seepage in the vertical plane, k disappears from the governing Laplace's equation and q_s becomes a linear function of k.

Following the previously described procedure, the following equations were obtained for the minimum seepage loss circular canal (Swamee and Kashyap 2001):

$$D^* = 1.540L \tag{7.1.10}$$

$$y^* = 0.300L \tag{7.1.11}$$

$$A^* = 0.256L^2 \tag{7.1.12}$$

$$q_s{}^* = 1.555kL \tag{7.1.13}$$

These values are also depicted in Table 7.1.

Example 7.1 Design a minimum seepage loss concrete-lined rectangular canal section for carrying a discharge of 50 m^3/s on a longitudinal slope of 0.0004. Consider drainage layer at large depth.

Solution For the design, $g = 9.79$ m/s^2, $\nu = 1.1 \times 10^{-6}$ m^2/s, and $\varepsilon = 1$ mm are adopted.

Using Eq. (5.1.9), $L = 9.890$ m. Using Table 7.1, the section shape coefficients are $k_b = 0.7986$, $k_y = 0.3178$, $k_A = 0.2538$, $k_V = 3.9402$, and $k_q = 2.0399$.

Using Eq. (5.1.18), $b^* = 0.7986 \times 9.890 = 7.898$ m; using Eq. (5.1.20), $y_n^* = 0.3178 \times 9.890 = 3.143$ m; using Eq. (5.1.22), $A^* = 0.2538 \times 9.890^2 = 24.824$ m^2; and using Eq. (5.1.23), $V^* = 3.9402 \times 50/9.890^2 = 2.013$ m/s, which is within the permissible limit. See Table 4.1.

Assuming that the lining is cracked and $k = 10^{-6}$ m/s, Eq. (7.1.9) results in the seepage loss $q_s^* = 2.0399 \times 10^{-6} \times 9.890 = 2.018 \times 10^{-5}$ m^2/s.

Example 7.2 Design a minimum seepage loss trapezoidal canal section for $Q = 250$ m^3/s and $S_o = 0.0001$.

Solution Following the steps similar to a rectangular section, $L = 23.954$ m. The section shape coefficients from Table 7.1 are $m^* = 0.5984$, $k_b = 0.5446$, $k_y = 0.3309$, $k_A = 0.2457$, $k_V = 4.0695$, and $k_q = 1.9227$.

Using Eqs. (5.1.18), (5.1.20), (5.1.21) and (5.1.22), $b^* = 0.5446 \times 23.954 = 13.045$ m; $y_n^* = 0.3309 \times 23.954 = 7.926$ m; $A^* = 0.2457 \times 23.954^2 = 141.00$ m^2; and $V^* = 4.0695 \times 250/23.954^2 = 1.773$ m/s, which is safe. Using Eq. (7.1.9) with $k = 10^{-6}$ m/s, $q_s^* = 1.9227 \times 10^{-6} \times 23.954 = 4.606 \times 10^{-5}$ m^2/s.

Power Law Channel: Swamee and Kashyap (2001) have conducted investigation on minimum seepage loss design of power law channels. Their results are summarized here. For $p \geq 1$, the design equations are as follows:

$$k_p^* = \frac{1.74p - 0.377}{1 + 0.695p} L^{(1-p)/p} \tag{7.1.14}$$

$$y_n^* = \frac{0.134 + 0.318p}{p} L \tag{7.1.15}$$

$$A^* = \frac{2\,(1 + 0.695p)\,(0.314 + 0.318p)^{(p+1)/p}}{(p+1)\,(1.74p - 0.377)\,p^{1/p}} L^2 \tag{7.1.16}$$

$$q_s^* = \left[\frac{78.04\,(33 + p)}{(25 + p)\,(50 + p)} + \frac{2.04\,(p - 1)}{50 + p}\right] kL \tag{7.1.17}$$

On the other hand, for $0.25 \leq p \leq 0.75$, the design equations are given as

$$k_p^* = (6 - 7p)\, L^{(1-p)/p} \tag{7.1.18}$$

$$y_n^* = (1.92 - 1.85p)\, L \tag{7.1.19}$$

$$A^* = \frac{2p(1.92 - 1.85p)^{(p+1)/p}}{(p+1)\,(6 - 7p)} L^2 \tag{7.1.20}$$

$$q_s^* = (6.25 - 5.7p)kL \tag{7.1.21}$$

For a parabolic section putting $p = 2$, Eqs. (7.1.14), (7.1.15), (7.1.16), and (7.1.17) simplify to

$$a^* = 0.5933L \tag{7.1.22}$$

$$y_n^* = 0.3850L \tag{7.1.23}$$

$$A^* = 0.2453L^2 \tag{7.1.24}$$

$$q_s^* = 1.9847kL \tag{7.1.25}$$

7.1.2 *Channels Having Drainage Layer at Shallow Depth*

For a triangular section, the seepage function is given by Eq. (2.4.16), whereas the uniform flow constraint is described by Eq. (7.1.5). As described in Sect. 7.1.1, the use of the minimization method results in the solution of the problem for a number of values of ε_*, ν_*, and d_* ($= d/L$) in the practical range. Making use of these optimal sections and adopting a procedure described in Sect. 5.1, the following empirical equations were obtained:

$$m^* = 1.2445\left[1 + \left(\frac{0.4826L}{d}\right)^{3.1847}\right]^{0.8232} \tag{7.1.26}$$

$$y_n^* = 0.4518L\left[1 + \left(\frac{0.5161L}{d}\right)^{0.3146}\right]^{-0..3177} \tag{7.1.27}$$

$$q_s^* = 2.0015kL\left[1 + \left(\frac{0.4385L}{d}\right)^{2.8994}\right]^{0.7238} \tag{7.1.28}$$

Following the above procedure, section shape equations for rectangular, and trapezoidal, were obtained. Generalized optimal equations for all these canal shapes are expressed as (Swamee et al. 2001)

$$m^* = k_{ms}\left[1 + \left(\frac{t_{md}L}{d}\right)^{r_{md}}\right]^{s_{md}} \tag{7.1.29}$$

Table 7.2 Coefficients and exponents for minimum seepage loss canals with drainage layer at shallow depth

Entity	Coefficient/exponent	Section shapes		
		Triangular	Rectangular	Trapezoidal
Side slope	k_{ms}	1.2445		0.5984
	t_{md}	0.4826		1.5156
	r_{md}	3.1847		3.5709
	s_{md}	0.8232		0.1077
Bed width	k_{bs}		0.7986	0.5447
	t_{bd}		0.4922	0.6042
	r_{bd}		2.5897	2.8821
	s_{bd}		0.5571	0.5129
Normal depth	k_{ys}	0.4518	0.3178	0.3309
	t_{yd}	0.5161	0.6293	0.6253
	r_{yd}	3.1465	2.8715	2.5376
	s_{yd}	0.3177	0.3125	0.3608
Seepage loss	k_{qs}	2.0015	2.0399	1.9227
	t_{qd}	0.4385	0.4417	0.4372
	r_{qd}	2.8994	2.2473	2.0318
	s_{qd}	0.7238	0.5807	0.6551

$$b^* = k_{bs} L \left[1 + \left(\frac{t_{bd} L}{d} \right)^{r_{bd}} \right]^{s_{bd}} \tag{7.1.30}$$

$$y_n^* = k_{ys} L \left[1 + \left(\frac{t_{yd} L}{d} \right)^{r_{yd}} \right]^{-s_{yd}} \tag{7.1.31}$$

$$q_s^* = k_{qs} k L \left[1 + \left(\frac{t_{qd} L}{d} \right)^{r_{qd}} \right]^{s_{qd}} \tag{7.1.32}$$

Table 7.2 lists section shape coefficients and exponents for the minimum seepage canal sections with drainage at shallow depth.

Swamee and Kashyap (2004) obtained the following empirical equations for circular section:

$$D^* = \frac{0.6L^2 + 1.078d^2}{L^2 + 0.7d^2} L \tag{7.1.33}$$

$$y_n^* = \frac{0.4L^{2.5} + 0.3d^{2.5}}{L^{2.5} + d^{2.5}} L \tag{7.1.34}$$

$$q_s^* = \frac{2.4L^4 + 6.22d^4}{L^4 + 4d^4} k L \tag{7.1.35}$$

On the other hand, Swamee and Kashyap (2004) found the following equations for a parabolic section:

$$a^* = \frac{0.3L^{10} + 0.415d^{10}}{L^{10} + 0.7d^{10}} L \tag{7.1.36}$$

$$y_n^* = \frac{0.486L^5 + 0.0847d^5}{L^5 + 0.22d^5} L \tag{7.1.37}$$

$$q_s^* = \frac{3.875L^{5.6} + 198.5d^{5.6}}{L^{5.6} + 100d^{5.6}} kL \tag{7.1.38}$$

The quantity of seepage being more, the drainage layer at shallow depth case is critical in designing canal of minimum seepage section. A perusal of Eqs. (7.1.29), (7.1.30), and (7.1.31) reveals that the minimum seepage section becomes wider and shallower as the canal bed approaches drainage layer. The minimum seepage section becomes impractical (strip) for $d < 0.01\,L$. For $L > d$, the canal dimensions are very sensitive. However, for $d \geq 3\,L$, the drainage layer could be assumed at very large depth. The seepage function becomes independent of depth of drainage layer and acquires a value corresponding to the drainage layer at infinity in a homogeneous porous medium of infinite extent and hence equations reduce to given in Sect. 7.1.1.

Example 7.3 Design a minimum seepage loss concrete-lined rectangular canal section for carrying a discharge of 50 m^3/s on a longitudinal slope of 0.0004. The canal passes through a stratum underlain by a highly pervious layer at a depth of 5 m.

Solution For the design, $g = 9.79$ m/s^2, $\nu = 1.007 \times 10^{-6}$ m^2/s (water at 20 °C), and $\varepsilon = 1$ mm (concrete lining) are adopted. Using Eq. (5.1.9), $L = 9.889$ m.

Using Table 7.2, the section shape coefficients and exponents for a rectangular section are $k_{bs} = 0.7986$, $t_{bd} = 0.4922$, $r_{bd} = 2.5897$, and $s_{bd} = 0.5571$; for normal depth, $k_{ys} = 0.3178$, $t_{yd} = 0.6293$, $r_{yd} = 2.8715$, and $s_{yd} = 0.3125$; and for seepage, $k_{qs} = 2.0399$, $t_{qd} = 0.4417$, $r_{qd} = 2.2473$, and $s_{qd} = 0.5807$.

Assuming a drainage layer at large depth, Eqs. (5.1.18) and (5.1.20) give $b^* = 0.7986 \times 9.889 = 7.897$ m and $y_n^* = 0.3178 \times 9.889 = 3.143$ m, respectively. Thus, $A^* = b^* y_n^* = 24.823$ m^2 and $V = 50/24.823 = 2.014$ m/s, which is safe. See Table 4.1. Further, assuming that the lining is cracked and $k = 10^{-6}$ m/s, Eq. (7.1.9) yields $q_s = 2.0399 \times 10^{-6} \times 9.890 = 2.0175 \times 10^{-5}$ m^2/s.

Using Eq. (7.1.30), for $d = 5$ m, $b^* = 0.7986 \times 1.4436 \times 9.889 = 11.400$ m, and using Eq. (7.1.31), $y_n^* = 0.3178 \times 0.7189 \times 9.889 = 2.259$ m. Therefore, $A = 11.400 \times 2.259 = 25.769$ m^2 and $V = 50/25.769 = 1.940$ m/s, which is within the permissible limit (Table 4.1). Further, using Eq. (7.1.32), the seepage loss $q_s = 2.0399 \times 1.3786 \times 10^{-6} \times 9.889 = 2.7813 \times 10^{-5}$ m^2/s. The design shows that the optimal section is influenced very much by the presence of a drainage layer at the shallow depth.

Example 7.4 Design a minimum seepage loss trapezoidal canal section for $Q = 250$ m/s, $S_o = 0.0001$, and $d = 7.5$ m.

Solution Following the steps similar to the rectangular section, $L = 23.950$ m. The section shape coefficients from Table 7.2 are for side slope, $k_{ms} = 0.5984$, $t_{md} = 1.5156$, $r_{md} = 3.5709$, and $s_{md} = 0.1077$; for bed width, $k_{bs} = 0.5446$, $t_{bd} = 0.6042$, $r_{bd} = 2.8821$, and $s_{bd} = 0.5129$; for normal depth, $k_{ys} = 0.3309$, $t_{yd} = 0.6253$, $r_{yd} = 2.5376$, and $s_{yd} = 0.3608$; and for seepage, $k_{qs} = 1.9227$, $t_{qd} = 0.4372$, $r_{qd} = 2.0318$, and $s_{qd} = 0.6551$.

Using Eqs. (7.1.29), (7.1.30) and (7.1.31), $m = 1.098$, $b^* = 37.042$ m, and $y_n^* = 3.972$ m. Therefore, $A = 164.454$ m^2 and $V = 250/164.454 = 1.520$ m/s, which is safe. Using Eq. (7.1.32) with $k = 10^{-6}$ m/s, $q_s = 1.9227 \times 2.0406 \times 10^{-6} \times 23.954 = 9.3983 \times 10^{-5}$ m^2/s. Assuming the drainage layer at large depth, the canal dimensions are $m = 0.598$, $b^* = 13.045$ m, and $y_n^* = 7.926$ m.

Example 7.5 Design a minimum seepage loss, concrete-lined circular canal section for carrying a discharge of 5 m^3/s on a longitudinal slope of 0.005. The canal passes through a stratum underlain by a highly pervious layer at a depth of 5 m.

Solution For the design, $g = 9.79$ m/s^2, $\nu = 1.1 \times 10^{-6}$ m^2/s (water at 20 °C), and $\varepsilon = 1$ mm (concrete lining) are adopted. Using Eq. (5.1.9), $L = 2.514$ m. Taking $d = 5$ m and using Eqs. (7.1.33) and (7.1.34), $D^* = 3.25$ m and $y_n^* = 0.758$ m, respectively. Further, assuming that the lining is cracked and $k = 10^{-6}$ m/s, Eq. (7.1.35) gave $q_s^* = 3.943 \times 10^{-6}$ m^2/s. Using Eqs. (7.1.10), (7.1.11), and (7.1.13), the corresponding figures for the drainage layer at an infinite depth are $D^* = 3.817$ m, $y_n^* = 0.754$ m, and $q_s^* = 3.909 \times 10^{-6}$ m^2/s.

Example 7.6 Design a minimum seepage loss, concrete-lined parabolic canal section for $Q = 50$ m^3/s, $S_o = 0.0005$, and $d = 7.5$ m.

Solution Following steps similar to those in the circular section, $L = 9.890$ m. With $d = 7.5$ m and using Eqs. (7.1.36) and (7.1.37), $a^* = 3.089$ m and $y_n^* = 4.735$ m, respectively. Considering the lining is cracked and $k = 10^{-6}$ m/s, Eq. (7.1.38) gave $q_s^* = 2.047 \times 10^{-5}$ m^2/s. Using Eqs. (7.1.22), (7.1.23), and (7.1.25), the corresponding figures for the drainage layer at an infinite depth are $a^* = 5.852$ m and $y_n^* = 3.808$ m, yielding $q_s^* = 1.963 \times 10^{-5}$ m^2/s.

7.2 Inclusion of Evaporation Loss

Denoting $q_{w*} = q_w/(k\lambda)$, $E_* = E/k$ and $T_* = T/\lambda$, and using Eq. (2.4.19), the design problem of total water loss canal section is written as

$$\text{Minimize} \quad q_{w*} = y_{n*}F_s + E_*T_* \tag{7.2.1}$$

subject to the equality constraint given by Eq. (7.1.2).

This constrained optimization problem was numerically solved by using the optimization process described in Sect. 5.1. Further, the optimization algorithm was applied on triangular, rectangular, and trapezoidal canal sections for a number of input variables varying in their practical ranges ($0 \le E \le 100k$). Analysis of these large numbers of optimal sections so obtained for all the three types of canal sections indicated that the linear dimensions are proportional to the length scale L. Further the analysis resulted in the following generalized equations for the optimal dimensions and the corresponding water loss for all the three canal sections (Swamee et al. 2002):

$$m^* = k_{ms}\left[1 + k_{me}\left(\frac{E}{k}\right)^{r_m}\right]^{s_m} \tag{7.2.2}$$

$$b^* = k_{bs}L\left[1 + k_{be}\left(\frac{E}{k}\right)^{r_b}\right]^{s_b} \tag{7.2.3}$$

$$y_n^* = k_{ys}L\left[1 + k_{ye}\left(\frac{E}{k}\right)^{r_y}\right]^{-s_y} \tag{7.2.4}$$

$$q_w^* = k_{qs}kL\left[1 + k_{qe}\left(\frac{E}{k}\right)^{r_q}\right]^{s_q} \tag{7.2.5}$$

The section shape coefficients and exponents appearing in Eqs. (7.2.2), (7.2.3), (7.2.4), and (7.2.5) for minimum water loss triangular, rectangular, and trapezoidal canal sections are listed Table 7.3.

Table 7.3 Coefficients and exponents for minimum water loss canals

		Section shapes		
Entity	Coefficients/exponents	Triangular	Rectangular	Trapezoidal
Side slope	k_{ms}	1.2466		0.5984
	k_{me}	0.4850		0.3106
	r_m	0.9020		1.0937
	s_m	1.1732		5.0000
Bed width	k_{bs}		0.7986	0.5446
	k_{be}		1.0717	0.1561
	r_b		0.9849	2.20409
	s_b		0.3798	0.1616
Normal depth	k_{ys}	0.4518	0.3178	0.3309
	k_{ye}	0.3895	0.5198	0.4987
	r_y	0.9286	0.8994	0.9998
	s_y	0.7114	0.6630	0.5828
Seepage loss	k_{qs}	2.0015	2.0399	1.9227
	k_{qe}	0.9084	0.5707	0.6249
	r_q	1.0126	0.9433	0.8849
	s_q	0.6241	0.6376	0.6846

For a given set of data, the use of Eqs. (7.2.2), (7.1.3), (7.1.4), and (7.2.5), along with Table 7.3, results in the optimal canal section. Section shape coefficients for water loss along with Eq. (7.2.5) give the minimum water loss from the optimal section.

Example 7.7 Design a minimum water loss concrete-lined rectangular canal section for carrying a discharge of 10 m^3/s on a longitudinal slope of 0.001. The canal lining has $\varepsilon = 1$ mm. Assume canal lining as cracked and having $k = 10^{-6}$ m/s. The maximum evaporation loss E was estimated as 2.5×10^{-6} m/s. The water temperature is 20 °C at which $\nu = 1.1 \times 10^{-6}$ m^2/s.

Solution Adopting $g = 9.79$ m/s^2 and using Eq. (5.1.9), $L = 4.471$ m. For a rectangular section, Table 7.3 gives the bed width parameters as $k_{bs} = 0.7986$, $k_{be} = 1.0717$, $r_b = 0.985$, and $s_b = 0.3798$. Using these parameters and Eq. (7.2.3), $b^* = 2.186$ m. Similarly, the normal depth parameters are $k_{ys} = 0.3178$, $k_{ye} = 0.5198$, $r_y = 0.8994$, and $s_y = 0.6630$. Using these parameters and Eq. (7.2.4), $y_n{}^* = 2.386$ m. Adopting $b = 2.19$ m and $y_n = 2.39$ m, $A = 5.234$ m^2 and $V = 10/5.234 = 1.91$ m/s, which is within permissible limit (Table 4.1). Using Eq. (7.2.1), the seepage loss $q_s = 1.032 \times 10^{-5}$ m^2/s and the evaporation loss $q_e = 5.475 \times 10^{-6}$ m^2/s. Summing up these losses, loss $q_w = 1.580 \times 10^{-5}$ m^2/s, while the direct use of Eq. (7.2.5) gives $q_w = 1.575 \times 10^{-5}$ m^2/s.

References

Sharma HD, Chawla AS (1975) Manual of canal lining, Technical report no.14. Central Board of Irrigation and Power, New Delhi

Swamee PK, Kashyap D (2001) Design of minimum seepage-loss nonpolygonal canal sections. J Irrig Drain Eng ASCE 127(2):113–117

Swamee PK, Kashyap D (2004) Design of minimum seepage-loss nonpolygonal canal section with drainage layer at shallow depth. J Irrig Drain Eng ASCE 130(2):166–170

Swamee PK, Mishra GC, Chahar BR (2000) Design of minimum seepage loss canal sections. J Irrig Drain Eng ASCE 126(1):28–32

Swamee PK, Mishra GC, Chahar BR (2001) Design of minimum seepage loss canal sections with drainage layer at shallow depth. J Irrig Drain Eng ASCE 127(5):287–294

Swamee PK, Mishra GC, Chahar BR (2002) Design of minimum water loss canal sections. J Hydraul Res 40(2):215–220

Chapter 8
Overall Minimum Cost Canal Sections

Abstract Design of a minimum cost canal section involves minimization of the sum of earthwork cost, cost of lining, and cost of water lost as seepage and evaporation subject to uniform flow condition in the canal. Essentially, it is a problem of minimization of a nonlinear objective function subject to a nonlinear equality constraint. This chapter highlights design equations for the least cost canal sections considering earthwork cost which may vary with depth of excavation, cost of lining, and cost of water lost as seepage and evaporation from irrigation canals of triangular, rectangular, and trapezoidal shapes passing through a stratum underlain by a drainage layer at shallow depth. Using nonlinear optimization technique on augmented function, generalized empirical equations and section shape coefficients have been obtained for the design of minimum overall cost canal sections of triangular, rectangular, and trapezoidal shapes. The optimal dimensions for any shape can be obtained from proposed equations along with tabulated section shape coefficients. The optimal design equations are in explicit form and result into optimal dimensions of a canal in single-step computations that avoid the trial and error method of canal design and overcome the complexity of the minimum cost design of canals by a constrained nonlinear optimization technique. The optimal design equations show that the optimal section becomes wider and shallower than the minimum area section due to additional cost of excavation with canal depth, while reverse is the case due to cost of water lost as evaporation. On the other hand, for increased lining cost and/or the excavation cost at ground level, the optimal canal section approaches to the minimum area section, while for increased cost of water lost as seepage, it approaches to the minimum seepage loss section. Design examples with sensitivity analysis demonstrate the simplicity of the proposed design equations.

Keywords Overall minimum cost canal section • Optimal sections • Minimum water loss section • Minimum seepage loss section • Minimum cost lined section • Minimum earthwork cost section • Minimum cost general section • Section shape coefficient • Explicit design equations • Design steps

A network of canals represents a major cost item in an irrigation project, and the economy of the canal network is vital. The maximum economy is achieved by minimizing the cost of the canals. Design of minimum cost irrigation canals

P.K. Swamee, B.R. Chahar, *Design of Canals*, Springer Transactions in Civil and Environmental Engineering, DOI 10.1007/978-81-322-2322-1_8

involves minimization of the sum of earthwork cost which varies with canal depth, cost of lining, and cost of water lost as seepage and evaporation subject to uniform flow condition in the canal. Such minimum cost canal design problem results in nonlinear objective function and nonlinear equality constraint making problem hard to solve analytically. In this chapter, generalized empirical equations and section shape coefficients for the design of minimum cost irrigation canal sections have been presented for triangular, rectangular, and trapezoidal canals.

8.1 Analytical Considerations

Using the overall cost function as given by Eq. (2.5.13), the problem of minimization of overall cost function can be reduced to the following nondimensional form:

$$\text{minimize } C_* = A_* + c_{r*}A_*\overline{y}_{n*} + c_{L*}P_* + c_{ws*}y_{n*}F_s + c_{wE*}T_* \tag{8.1.1}$$

subject to the constraint given by Eq. (7.1.2)
where

$$\overline{y}_{n*} = \overline{y}_n\left(gS_o/Q^2\right)^{0.2} \tag{8.1.2}$$

$$P_* = P\left(gS_o/Q^2\right)^{0.2} \tag{8.1.3}$$

$$y_{n^*} = y_n\left(gS_o/Q^2\right)^{0.2} \tag{8.1.4}$$

$$T_* = T\left(gS_o/Q^2\right)^{0.2} \tag{8.1.5}$$

$$c_{ws*} = \frac{c_{ws}}{c_e}\left(\frac{gS_o}{Q^2}\right)^{0.2} \tag{8.1.6}$$

$$c_{wE*} = \frac{c_{wE}}{c_e}\left(\frac{gS_o}{Q^2}\right)^{0.2} \tag{8.1.7}$$

Using the exterior penalty function method as described in Sect. 5.1 and varying in the ranges of $10^{-6} \le \varepsilon_* \le 10^{-3}$,$10^{-7} \le \nu_* \le 10^{-5}$, $0 \le c_{L*} \le \infty$, $0 \le c_{r*} \le 50c_{L*}$, $0 \le c_{ws*} \le \infty$, and $0 \le c_{wE*} \le c_{ws*}$, optimum sections were found out. Analysis of a large number of optimal sections so obtained for triangular, rectangular, and trapezoidal canal sections resulted in the following generalized empirical equations (Swamee et al. 2000):

$$m^* = k_{m0}\frac{k_{mL}c_L + k_{ms1}c_{ws}\left(1 + k_{md}L/d\right) + c_eL + k_{mr}c_rL^2}{k_{mL}c_L + k_{ms2}c_{ws} + k_{mE}c_{wE} + c_eL} \tag{8.1.8}$$

$$b^* = k_{b0}\frac{k_{bL}c_L + k_{bs1}c_{ws}\,(1 + k_{bd}\,L/d) + c_eL + k_{br}c_rL^2}{k_{bL}c_L + k_{bs2}c_{ws} + k_{bE}c_{wE} + c_eL}L \quad (8.1.9)$$

$$y_n^* = k_{y0}\frac{k_{yL}c_L + k_{ys2}c_{ws} + k_{yE}c_{wE} + c_eL}{k_{yL}c_L + k_{ys1}c_{ws}\left(1 + k_{yd}\,L/d\right) + c_eL + k_{yr}c_rL^2}L \quad (8.1.10)$$

$$C^* = [k_{cL}c_L + k_{cs}c_{ws}\,(1 + k_{cd}\,L/d) + k_{cE}c_{wE}]\,L + k_{ce}c_eL^2 + k_{cr}c_rL^3 \quad (8.1.11)$$

Table 8.1 lists the section shape coefficients.

These equations are applicable for full range of lining cost ($0 \le c_{L*} \le \infty$) and cost of water lost due to seepage in the absence of drainage layer ($0 \le c_{ws*} \le \infty$).

Table 8.1 Properties of optimal canal sections

Entity	Section shape coefficients	Section shapes Triangular	Rectangular	Trapezoidal
(1)	(2)	(3)	(4)	(5)
Side slope	k_{m0}	1.0000		0.5774
	k_{mL}	15.049		14.2772
	k_{mr}	0.304		0.216
	k_{ms1}	16.756		23.494
	k_{ms2}	13.441		22.668
	k_{md}	0.125		0.124
	k_{mE}	8.331		32.189
Bed width	k_{b0}		0.711	0.4341
	k_{bL}		15.028	14.243
	k_{br}		0.320	0.348
	k_{ms1}		18.283	18.086
	k_{ms2}		16.286	14.416
	k_{bd}		0.111	0.130
	k_{mE}		5.543	0.288
Normal depth	k_{y0}	0.503	0.356	0.376
	k_{yL}	15.039	15.023	14.227
	k_{yr}	0.140	0.307	0.2233
	k_{ys1}	16.245	18.737	16.910
	k_{ys2}	14.590	16.741	14.885
	k_{yd}	0.054	0.097	0.112
	k_{yE}	4.036	5.624	4.036
Cost	k_{co}	0.2530	0.253	0.245
	k_{cL}	1.425	1.424	1.303
	k_{cr}	0.040	0.040	0.037
	k_{cs}	2.002	2.040	1.923
	k_{cd}	0.142	0.134	0.161
	k_{cE}	0.989	0.686	0.820

Also when earthwork cost and/or cost of water lost due to evaporation and seepage with drainage layer is considered, then these equations are applicable if $0 \leq c_{r*} \leq 1.0$ or $0 \leq c_{r*} \leq 50\ c_{L*}$, $0 \leq c_{wE*} \leq 0.2$ or $0 \leq c_{wE*} \leq c_{ws*}$, and $d_* \geq 0.25$. In practical cases, these ranges are commonly satisfied. If earthwork cost is more important with $c_{r*} \geq 1.0$ and $c_{r*} \geq 50\ c_{L*}$, it is better to use minimum earthwork cost design equations of Chap. 6. Similarly, equations presented in Chap. 7 should be used for shallow drainage layer and evaporation cases with $c_{wE*} \geq 0.2$ and $c_{wE*} \geq c_{ws*}$ and $d_* \leq 0.25$.

8.2 Particular Cases

Equations (8.1.8) through (8.1.11) are the general equations for optimal canal design. Here, two particular cases are of special interest. These are minimum cost lined sections and minimum water loss sections. These sections are described below.

8.2.1 *Minimum Cost Lined Sections*

When the cost of water is not a limiting factor, $c_{ws} = c_{wE} = 0$. In such a case, Eqs. (8.1.8), (8.1.9), (8.1.10), and (8.1.11) reduce to

$$m^* = k_{m0} \frac{k_{mL} c_L + c_e L + k_{mr} c_r L^2}{k_{mL} c_L + c_e L} \tag{8.2.1}$$

$$b^* = k_{b0} \frac{k_{bL} c_L + c_e L + k_{br} c_r L^2}{k_{bL} c_L + c_e L} L \tag{8.2.2}$$

$$y_n^* = k_{y0} \frac{k_{yL} c_L + c_e L}{k_{yL} c_L + c_e L + k_{yr} c_r L^2} L \tag{8.2.3}$$

$$C^* = k_{cL} c_L L + k_{ce} c_e L^2 + k_{cr} c_r L^3 \tag{8.2.4}$$

It has to be noted that this particular case has been investigated in Chap. 6. Furthermore, if $c_r = 0$ in Eqs. (8.2.1) through (8.2.4), equations of minimum area section (Chap. 5) are generated as $m^* = k_{m0}$; $b^* = k_{b0} L$; $y_n^* = k_{y0} L$ and $C^* = k_{cL} c_L L + k_{ce} c_e L^2$.

8.2.2 *Minimum Water Loss Sections*

Considering $c_e = c_r = c_L = 0$ in Eqs. (8.1.8) through (8.1.11), equations of minimum water loss section are generated. Thus,

$$m^* = k_{m0} k_{ms1} \frac{1 + k_{md} L/d}{k_{ms2} + k_{mE} E/k} \tag{8.2.5}$$

$$b^* = k_{b0} k_{bs1} \frac{1 + k_{bd} L/d}{k_{bs2} + k_{bE} E/k} L \tag{8.2.6}$$

$$y_n^* = \frac{k_{y0}}{k_{ys1}} \frac{k_{ys2} + k_{yE2} E/k}{1 + k_{yd} L/d} L \tag{8.2.7}$$

$$q_w^* = k_{cs} k \left(1 + k_{cd} L/d\right) L + k_{cE} E L \tag{8.2.8}$$

where using Eqs. (2.5.14) and (2.5.15), c_{wE}/c_{ws} is substituted by E/k. It may be seen that the functional form of Eqs. (8.2.5) through (8.2.8) is much simpler than their counterpart described in Chap. 7.

A perusal of Eqs. (8.1.8), (8.1.9), (8.1.10), and (8.1.11) with Table 8.1 indicates that for $c_r = 0$, $c_{ws} = 0$, and $c_{wE} = 0$, the optimal section is the minimum area section. However, with increase in c_r, the canal section becomes wider and shallow, while the reverse is the case with increase in c_{wE}. On the other hand, with increase in c_e and/or c_L, the canal section approaches the corresponding minimum area section, while with increase in c_{ws}, it approaches the minimum seepage loss section (Swamee et al. 2000).

8.3 Design Steps

Design equations for different cases have been presented in Chaps. 5, 6, 7, and 8. For all these cases, an optimal section is designed by adopting the following steps:

- For a given set of data (g, Q, S_o, ε, and ν), find L using Eq. (5.1.9).
- List the section shape coefficients from the appropriate table (say, Table 8.1) for the desired section shape (say, trapezoidal) and minimization case (say, minimum overall cost section).
- Use the appropriate optimal design equations [say, Eqs. (8.1.8), (8.1.9), and (8.1.10)] along with section shape coefficients and cost results in the optimal canal section.
- For the designed section, the average flow velocity V can be obtained by continuity of Eq. (3.1.1). This velocity should be greater than the non-silting velocity but less than the limiting velocity V_L (see Table 4.2).
- If V is greater than V_L, use a canal route with a smaller bed slope, adopt a flatter bed slope with drops, build two smaller canals, select a superior lining material having higher limiting velocity, use a lining material with a larger roughness, or opt for a nonoptimal section.
- Find the minimum cost for the optimal section using the appropriate optimal cost equation [say Eq. (8.1.11)] for the case in hand. Alternatively, the cost of the section can be obtained from the cost function once the section dimensions are decided.

Example 8.1 Design a concrete lined trapezoidal canal section for carrying a discharge of 250 m^3/s on a longitudinal bed slope of 0.0001. The canal passes through a stratum of ordinary soil for which $c_e/c_r = 7$ m, $k = 10^{-6}$ m/s, and depth of drainage layer = 7.5 m. Further, it is proposed to provide concrete lining with $c_L/c_e = 12$ m. The climatic condition of the canal area is such that the maximum evaporation loss was estimated as 2.0×10^{-7} m/s (17.28 mm/day). Take rate of interest = 0.04 ₹/₹/year and $c_w/c_e = 0.013$.

Solution For the design, $g = 9.79$ m/s^2, $\nu = 1.007 \times 10^{-6}$ m^2/s, and $\varepsilon = 1$ mm (for smooth finished concrete lining) have been adopted. Using Eq. (5.1.9), $L = 23.950$ m.

Minimum Cost General Section

For a trapezoidal section, Table 8.1 gives section shape coefficients: for side slope, $k_{m0} = 0.57735$, $k_{mr} = 0.2163$, $k_{mL} = 14.277$, $k_{ms1} = 23.494$, $k_{ms2} = 22.668$, $k_{md} = 0.1235$, and $k_{mE} = 32.189$; for bed width, $k_{b0} = 0.43407$, $k_{br} = 0.3484$, $k_{bL} = 14.243$, $k_{bs1} = 18.086$, $k_{bs2} = 14.416$, $k_{bd} = 0.1302$, and $k_{bE} = 0.2878$; for normal depth, $k_{y0} = 0.37592$, $k_{yr} = 0.2233$, $k_{yL} = 14.227$, $k_{ys1} = 16.910$, $k_{ys2} = 14.886$, $k_{yd} = 0.1122$, and $k_{yE} = 4.0362$; and for cost, $k_{ce} = 0.24476$, $k_{cr} = 0.03723$, $k_{cL} = 1.3037$, $k_{cs} = 1.9227$, $k_{cd} = 0.1814$, and $k_{cE} = 0.8194$.

Using Eqs. (8.1.8), (8.1.9), and (8.1.10) with these coefficients, $m^* = 0.642$, $b^* = 14.718$ m, and $y_n^* = 7.138$ m. Further, Eq. (8.1.11) gives the earthwork cost per meter $C_e = 140.39c_e + 73.07c_e = 213.46c_e$, the lining cost per meter $C_L = 374.68c_e$, and the cost of water lost per meter $C_w = 715.00c_e + 40.26c_e = 755.26c_e$. Thus, the canal cost per meter $= C_e + C_L + C_w = 1343.4c_e$. It can be seen that the cost of lining and the cost of water lost as seepage share the major portion of the total cost.

Minimum Cost Lined Section

If cost of the water is not taken into account, the canal design parameters, using Eqs. (8.2.1), (8.2.2), (8.2.3), and (8.2.4), are $m^* = 0.630$, $b^* = 11.920$ m, $y_n^* = 8.23$ m, and $C^* = 588.14c_e$.

Minimum Earthwork Cost Section

For a trapezoidal section, Table 6.2 gives section shape coefficients: for side slope, $k_{m0} = 0.57735$, $t_{mr} = 10.00$, $r_{mr} = 1.2586$, and $s_{mr} = 0.08069$; for bed width, $k_{b0} = 0.43407$, $t_{br} = 0.3195$, $r_{br} = 0.9342$, and $s_{br} = 1.0712$; for normal depth, $k_{y0} = 0.37592$, $t_{yr} = 0.2088$, $r_{yr} = 0.8123$, and $s_{yr} = 1.0280$; and for cost, $k_{ce} = 0.24476$, $t_{cr} = 0.2277$, $r_{cr} = 0.9544$, and $s_{cr} = 0.6855$.

With $c_{r*} = 5.199$ and these coefficients, Eqs. (6.2.1), (6.2.2), (6.2.3), and (6.2.4) yield $m^* = 0.789$, $b^* = 21.939$ m, and $y_n^* = 5.673$ m. Further, Eq. (6.2.5) gives cost per meter length of canal $C^* = 204.96c_e$. These dimensions result to $A = y_n(b + my_n) = 149.852$ m^2. Thus, $V = 250/149.852 = 1.668$ m/s.

On the other hand, the corresponding design values obtained for the depth-independent earthwork cost (minimum area) section are $m^* = 0.57735$, $b^* = 10.396$ m, $y_n^* = 9.003$ m, $A = 140.392$ m^2, and $V = 1.781$ m/s. The cost canal section with these dimensions comes out $C^* = 220.64c_e$. The design reveals that there is 111 % increase in the bed width, 37 % decrease in the normal depth, and 36.7 % increase in the side slope of the canal section when the earthwork cost as a function of depth is considered.

Minimum Seepage Loss Section

$d_* = 0.206$. The section shape coefficients from Table 7.2: $k_{ms} = 0.59836$, $t_{md} = 1.5156$, $r_{md} = 3.5709$, $s_{md} = 0.1077$, $k_{bs} = 0.54458$, $t_{bd} = 0.6042$, $r_{bd} = 2.8821$, $s_{bd} = 0.5129$, $k_{ys} = 0.3309$, $t_{yd} = 0.6253$, $r_{yd} = 2.5376$, $s_{yd} = 0.3608$, $k_{qs} = 1.92272$, $t_{qd} = 0.4372$, $r_{qd} = 2.0318$, and $s_{qd} = 0.6551$.

Using Eqs. (7.1.29), (7.1.30), and (7.1.31), $m^* = 1.098$, $b^* = 37.042$ m, and $y_n^* = 3.972$ m. Therefore, $A = 164.454$ m^2 and $V = 1.520$ m/s, which is safe. Assuming a cracked lining and using Eq. (7.1.32) with $k = 10^{-6}$ m/s, $q_s^* = 1.92272 \times 2.0406 \times 10^{-6} \times 23.950 = 9.3983 \times 10^{-5}$ m^2/s.

Assuming the drainage layer at large depth and using Table 7.1, the canal dimensions are $m^* = 0.598$, $b^* = 13.045$ m, and $y_n^* = 7.926$ m.

Minimum Water Loss Section

$E = 0.2k$. The section shape coefficients from Table 7.3: $k_{ms} = 0.59836$, $t_{m\mathrm{E}} = 0.3106$, $r_{m\mathrm{E}} = 1.0937$, $s_{m\mathrm{E}} = 5.000$, $k_{bs} = 0.54458$, $t_{b\mathrm{E}} = 0.1561$, $r_{b\mathrm{E}} = 2.2409$, $s_{b\mathrm{E}} = 0.1616$, $k_{ys} = 0.3309$, $t_{y\mathrm{E}} = 0.4987$, $r_{y\mathrm{E}} = 0.9998$, $s_{y\mathrm{E}} = 0.5828$, $k_{qs} = 1.92272$, $t_{q\mathrm{E}} = 0.6249$, $r_{q\mathrm{E}} = 0.8849$, and $s_{q\mathrm{E}} = 0.6846$.

Assuming drainage layer at a large depth and using Eqs. (7.2.2), (7.2.3), and (7.2.4), $m^* = 0.461$, $b^* = 13.034$ m, and $y_n^* = 8.377$ m. Therefore, $A = 141.536$ m^2 and $V = 1.766$ m/s. Further, using Eq. (7.2.4), $q_w^* = 1.92272 \times 1.1007 \times 10^{-6} \times 23.950 = 5.068 \times 10^{-5}$ m^2/s.

Sensitivity of Optimal Design

Assuming the drainage layer at very large depth, the design variables in the general minimum cost section become $m^* = 0.531$, $b^* = 12.382$ m, $y_n^* = 8.286$ m, and $C^* = 1{,}100.7c_e$. For bed width ranging from 0 m to 40 m, and side slope ranging from 0 to 5, the normal depths and costs have been calculated. Figure 8.1a shows the variation of cost with b and m. The cost per meter length of canal is least ($C = 1{,}100.7c_e$) for a trapezoidal section with side slope 0.531 and bed width 12.382 m. It can be seen that the optimum is less sensitive to the increase in bed width and more sensitive to the decrease in b. The optimum for rectangular section ($m = 0$) occurs at $b = 18.62$ m ($C = 1{,}156.5c_e$), and the cost is highly sensitive to decrease in bed width as seen in Fig. 8.1a. For the same data, the triangular section ($b = 0$) has minimum cost ($C = 1{,}160.3c_e$) at side slope 1.116. Figure 8.1a also shows that for the design data, the rectangular section is more economical for

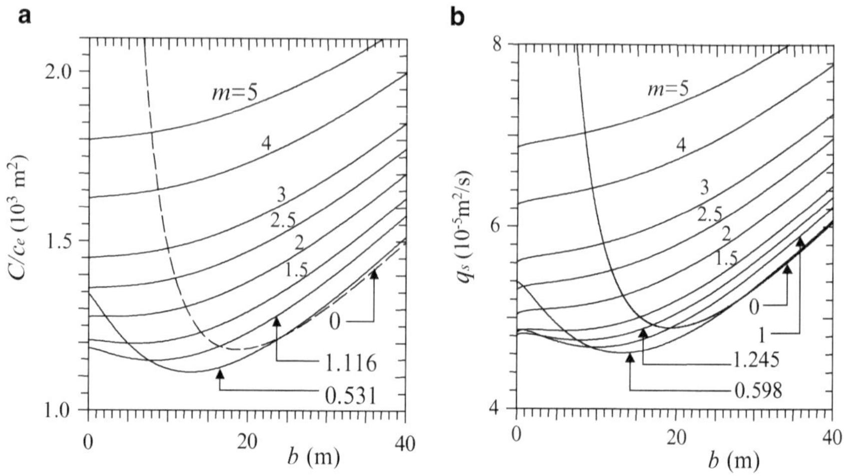

Fig. 8.1 Sensitivity of the optimal design. (**a**) Minimum cost general section. (**b**) Minimum seepage loss section

$b > 24$ m, while the triangular section is more economical for higher side slopes ($m > 2$), and the trapezoidal section is the most economical section otherwise.

A similar sensitivity of the minimum seepage loss section, assuming the drainage layer at infinity, has also been carried out as shown in Fig. 8.1b. It can be observed that the trend of the sensitivity is the same as the pervious case. Figure 8.1b shows that the seepage loss from a trapezoidal section with side slope of 0.598 and bed width 13.045 m is the global minimum. Furthermore, for $m > 1.75$, the optimum shifts to $b = 0$ (triangular section). However, the optimum for a rectangular section ($m = 0$) is highly sensitive to a decrease in bed width.

Example 8.2 Design a triangular canal section for $Q = 50$ m³/s, $S_o = 0.0004$, $d = 5$ m, and other data being the same as in Example 8.1.

Solution For the given data, $L = 9.889$ m.

Minimum Cost General Section

Using Eqs. (8.1.8) and (8.1.10) for a triangular section with section shape coefficients from Table 8.1 gives $m^* = 1.184$ and $y_n^* = 4.599$ m. Further, Eq. (8.1.11) gives $C_e = 24.74c_e + 5.48c_e = 30.22c_e$, $C_L = 169.09c_e$, $C_w = 260.15c_e + 20.06c_e = 280.21c_e$, and hence $C = 479.52c_e$. Adopting $m = 1.2$ and $y_n = 4.6$ m, $A = 25.392$ m^2 and $V = 1.969$ m/s.

Minimum Cost Lined Section

The optimal dimensions for a triangular canal section for this case, using Eqs. (8.2.1) and (8.2.3), are $m^* = 1.022$ and ${y_n}^* = 4.924$ m, with corresponding cost per unit length of the canal $C = 199.31 c_e$.

Minimum Earthwork Cost Section

Using the section shape coefficients and exponents for the triangular section from Table 6.2 and design Eqs. (6.2.1), (6.2.4), and (6.2.5), $m^* = 1.373$, ${y_n}^* = 4.254$ m, and $C^* = 32.55 c_e$. Thus, $A = m^* \times {y_n}^* \times y_n = 24.847$ m^2 and $V = 50/24.847 = 2.012$ m/s. For a minimum area triangular section, $m^* = 1.0$, ${y_n}^* = 4.974$ m, and $C^* = 32.55 c_e$. Thus, $A = 24.743$ m^2.

Minimum Seepage Loss Section

Using Eqs. (7.1.29), (7.1.31) and (7.1.32) along with Table 7.2 for a triangular section yields $m^* = 2.076$, ${y_n}^* = 3.547$ m, and ${q_s}^* = 2.859 \times 10^{-5}$ m^2/s. Adopting $m = 2.1$ and $y_n = 3.55$ m, $A = 26.46$ m^2 and $V = 1.89$ m/s. Assuming homogeneous medium of infinite extent or not considering the drainage layer and using Eqs. (7.1.6), (7.1.7) and (7.1.8), the optimal parameters are $m^* = 1.2445$, ${y_n}^* = 4.468$ m, and ${q_s}^* = 1.979 \times 10^{-5}$ m^2/s. The actual seepage loss for these dimensions due to the existence of the drainage layer, using Eq. (2.4.1) along with Eq. (2.4.16), comes out 3.457×10^{-5} m^2/s, which is 20.9 % more than the optimal design considering the drainage layer at a shallow depth.

Minimum Water Loss Section

For a triangular section, Table 7.3 gives the side slope parameters as $k_{ms} = 1.2445$, $k_{me} = 0.4850$, $r_m = 0.9020$, and $s_m = 1.1732$. Using these parameters and Eq. (7.2.2), $m^* = 1.096$. Similarly, the normal depth parameters are $k_{ys} = 0.45177$, $k_{ye} = 0.3895$, $r_y = 0.9286$, and $s_y = 0.7114$. Using Eq. (7.2.4) with these parameters, ${y_n}^* = 4.742$ m. Using Eq. (7.2.5) with water loss parameters yields the water loss ${q_E}^* = 2.192 \times 10^{-5}$ m^2/s.

Comparison Among Optimal Designs

The optimal sections resulted for the data of Examples 8.1 and 8.2 for different minimization cases and for three shapes of a canal have been compared in Table 8.2. The table shows that there is considerable difference in canal dimensions for different minimization cases. For example, the optimal rectangular section in Example 8.2 for the drainage layer at shallow depth results in 28.11 % less y_n, 44.36 % more b, and 12.03 % less seepage than the optimal section considering drainage layer at large depth. A similar comparison can be made between two different optimal design cases.

Table 8.2 Comparison among canal shapes and dimensions

		Section shapes					
		Triangular		Rectangular		Trapezoidal	
Design case	Parameter	Ex. 1	Ex. 2	Ex. 1	Ex. 2	Ex. 1	Ex. 2
Minimum area section	m	1.00	1.00			0.5774	0.5774
	b			17.037	7.035	10.396	4.293
	y_n	12.047	4.974	8.519	3.517	9.003	3.718
	A	145.13	24.74	145.13	24.74	140.39	23.94
Minimum earthwork cost section	m	1.896	1.409			0.789	0.724
	b			30.070	9.677	21.939	6.350
	y_n	9.072	4.267	5.113	2.617	5.673	2.893
	C/c_e	216.72	29.86	211.21	29.93	204.95	28.90
Minimum seepage section drainage layer at infinite	m	1.2445	1.2445			0.5984	0.5984
	b			19.126	7.897	13.045	5.385
	y_n	10.820	4.468	7.611	3.143	7.926	3.272
	q_s (10^{-5})	4.794	1.979	4.886	2.017	4.605	1.901
Min. seepage section drain. layer at finite depth	m	4.654	2.076			1.098	0.915
	b			42.687	11.400	37.042	8.914
	y_n	6.184	3.547	3.911	2.259	3.972	2.371
	q_s (10^{-5})	12.247	2.859	10.162	2.781	9.398	2.735
Minimum water loss section	m	1.097	1.097			0.461	0.461
	b			17.737	7.324	13.034	5.382
	y_n	11.485	4.742	8.216	3.392	8.377	3.459
	q_w (10^{-5})	5.310	2.192	5.267	2.175	5.068	2.093
Minimum cost lined section	m	1.122	1.022			0.630	0.587
	b			19.225	7.200	11.920	4.408
	y_n	11.408	4.924	7.586	3.440	8.230	3.654
	C/c_e	632.46	199.31	632.13	199.20	588.14	183.79
Minimum cost general section	m	1.307	1.184			0.642	0.583
	b			21.572	8.080	14.718	5.462
	y_n	10.617	4.599	6.867	3.103	7.138	3.209
	C/c_e	1396.1	479.52	1380.9	474.73	1343.4	457.50

Example 8.3 Design a rectangular canal for Example 8.1 with $k = 5 \times 10^{-6}$ m/s.

Solution

Minimum Cost General Section

$L = 23.950$ m. Using $c_{ws} = 3.156 \times 10^7 k\ c_w/r$ and $c_{wE} = 3.156 \times 10^7 E\ c_w/r$ gave $c_{ws} = 3945.0c_w$ and $c_{wE} = 157.8c_w$, respectively. Thus, $c_{r*} = 5.200$, $c_{L*} = 0.330$, $c_{ws*} = 1.373$, $c_{wE*} = 0.055$, and $d_* = 0.2$, which are well within the application range of design equations (8.1.8) through (8.1.11).

For a rectangular section, Table 8.1 gives section shape coefficients: for bed width, $k_{be} = 0.71136$, $k_{br} = 0.3201$, $k_{b\mathrm{L}} = 15.028$, $k_{bs1} = 18.283$, $k_{bs2} = 16.286$, $k_{bd} = 0.1111$, and $k_{b\mathrm{E}} = 5.5434$; for normal depth, $k_{ye} = 0.35568$, $k_{yr} = 0.3066$, $k_{y\mathrm{L}} = 15.023$, $k_{ys1} = 18.737$, $k_{ys2} = 16.741$, $k_{yd} = 0.0964$, and $k_{y\mathrm{E}} = 5.6244$; and for cost, $k_{ce} = 0.25302$, $k_{cr} = 0.03961$, $k_{c\mathrm{L}} = 1.4240$, $k_{cs} = 2.0399$, $k_{cd} = 0.1337$, and $k_{c\mathrm{E}} = 0.6856$.

Using Eqs. (8.1.9) and (8.1.10) with these coefficients, $b^* = 24.306$ m and $y_n^* = 6.164$ m. Further, Eq. (8.1.11) gives the canal cost per meter $C^* = 77.74c_e + 145.13c_e + 409.26c_e + 3485.72c_e + 32.84c_e = 4150.69c_e$. These dimensions yield $A = b\ y_n = 149.82\ \mathrm{m}^2$. Thus, $V = 250.0/149.82 = 1.669$ m/s, which is within the permissible limit.

For a trapezoidal section, the corresponding parameters are $m^* = 0.767$, $b^* = 17.065$ m, $y_n^* = 6.264$ m, $A = 137.0\ \mathrm{m}^2$, and $C^* = 4{,}112.83c_e$.

In the absence of a drainage layer or when a drainage depth is at a very large depth, the optimal canal dimensions would be (a) for a rectangular section, $b^* = 18.940$ m and $y_n^* = 7.688$ m, and (b) for a trapezoidal section, $m^* = 0.575$, $b^* = 12.795$ m, and $y_n^* = 8.055$ m.

Minimum Cost Lined Section

If cost of the water is not taken into account, the canal design parameters, for a rectangular section, are $b^* = 19.225$ m, $y_n^* = 7.586$ m, and $C^* = 632.13c_e$. On the other hand, a trapezoidal section with $m^* = 0.630$, $b^* = 11.920$ m, $y_n^* = 8.23$ m, and $C^* = 588.14c_e$ will result into a minimum cost lined section.

Minimum Seepage Loss Section

Assuming the drainage layer at a large depth, the canal dimensions for a rectangular section are $b^* = 19.126$ m and $y_n^* = 7.611$ m and for a trapezoidal section are $m^* = 0.598$, $b^* = 13.045$ m, and $y_n^* = 7.926$ m.

Reference

Swamee PK, Mishra GC, Chahar BR (2000) A comprehensive design of minimum cost irrigation canal sections. J Irrig Drain Eng ASCE 126(5):322–327

Chapter 9
Design of Canal Transitions

Abstract A canal from source to destination may be of several hundred kilometers. The discharge in the canal varies along the length due to diversion and losses; therefore, a reduced canal section matching with the discharge is adopted. Canal section may also change at flumes, siphons, and aqueducts. A canal transition involving an expansion or contraction of the section is required whenever there is change in the canal section. A transition is a structure of short length; thus, the cost aspect of transitions is not considered in their design. This chapter gives design procedure for both contraction and expansion transitions.

Keywords Canal transition • Expansion transition • Contraction transition • Transition length • Energy loss • Optimal control problem • Optimal sections • Gradually varied flow

A canal transition involving an expansion and contraction of the section is a common feature of many hydraulic structures such as flumes, siphons, and aqueducts. For example, before a canal passes through an aqueduct, it is first reduced in cross section, and then it crosses the hydraulic structure with contracted section. Subsequently, the section is expanded to its original shape. Thus, the crossing involves both contraction and expansion transitions. Whereas in a contraction transition there is little energy loss, on the other hand, the head loss in an expansion transition is substantial. Thus, much attention has to be paid in the design of an expansion transition.

A general expansion transition design problem involves designing a transition from a rectangular channel of bed width b_0 to an expanded trapezoidal channel of bed width b_L and side slope m_L in a transition length L. The bed level also transitions from zero at the inlet to z_L at the outlet. (See Fig. 9.1). A rectangular transition is a limiting case when $m_L = 0$. On the other hand, a general contraction transition design problem involves designing a transition from a trapezoidal channel of bed width b_0 to a contracted rectangular channel of bed width b_L and side slope m_L in a transition length L. The design of such structures involves determination of a transition shape such that the energy loss incurred is minimum.

P.K. Swamee, B.R. Chahar, *Design of Canals*, Springer Transactions in Civil and Environmental Engineering, DOI 10.1007/978-81-322-2322-1_9

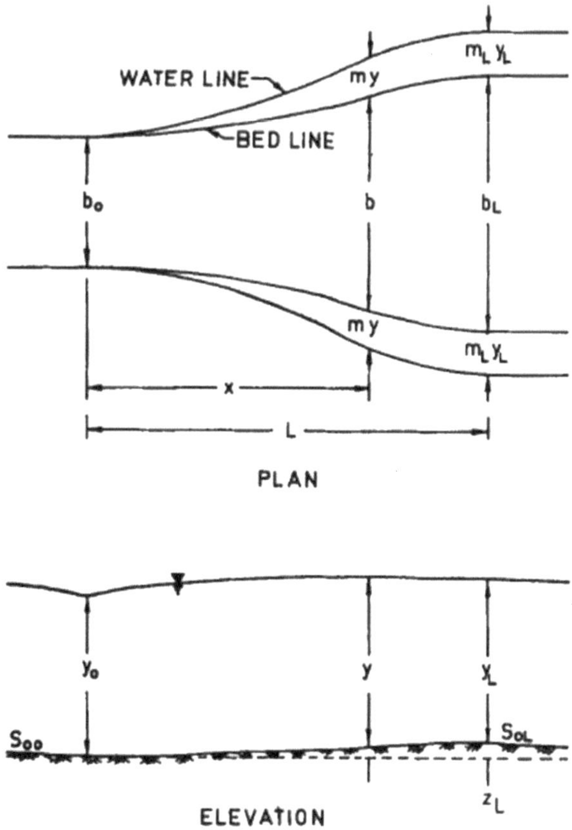

Fig. 9.1 Expansion transition

9.1 Expansion Transitions

The available information regarding design of the transition shape is mostly empirical. Considering flow depth to be constant and average velocity V varying linearly, Hartley et al. (1940) gave the following equation for transition of a rectangular to rectangular expansive transition:

$$b = \left[b_0^{-1} - \left(b_0^{-1} - b_L^{-1}\right)\, \xi\right]^{-1} \tag{9.1.1}$$

where $b =$ bed width at a distance x from inlet of the transition, b_0 and $b_L =$ the width of channel before and after the transition, respectively, and $\xi = x/L$.

In an expanding transition as the head loss due to surface resistance is small, it can be neglected. Thus, the total head loss can be written as the form loss alone. Applying the momentum and energy equations for an abrupt expansion, in a rectangular channel, Henderson (1966) gave the following equation for form loss:

$$h_L = \frac{V_0^2}{2g}\left[\left(1 - \frac{b_0}{b_1}\right)^2 + \frac{2\mathbf{F}_0^2 b_0^3}{b_L^4}(b_L - b_0)\right] \tag{9.1.2}$$

where $\mathbf{F}_0 = Q/\left(b_0 y_{n0}\sqrt{g y_{n0}}\right)$. Considering an elementary length dx of an expanding channel, and putting $b_0 = b,\ y_{n0} = y$ and $b_L = b + db,$ and further considering $db \to 0$, Eq. (9.1.2) reduces to

$$dh_L = \frac{Q^4}{g^2 b^5 y^5} db \tag{9.1.3}$$

where dh_L = the elementary head loss in the length dx. The head loss in the transition, thus, can be written as

$$h_L = \int_0^L \frac{Q^4}{g^2 b^5 y^5} w dx \tag{9.1.4}$$

where

$$\frac{db}{dx} = w \tag{9.1.5}$$

The water surface profile is given by the following generalization of the differential equation of gradually varied flow:

$$\frac{dy}{dx} = \frac{S_o + \frac{Q^2}{gb^3y^2}\left(1 - \frac{Q^2}{gb^3y^2}\right) w}{1 - \frac{Q^2}{gb^3y^2}} \tag{9.1.6}$$

Denoting $B = (b - b_0)/(b_L - b_0)$, $\mathbf{G} = Q/\left(b_0^2\sqrt{g b_0}\right)$, $r = b_L b_0$, $\eta = y/b_0$, and $H_L = h_L/b_0$, Eq. (9.1.6) is written in the following nondimensional form (see Swamee and Basak 1991):

$$H_L = \int_0^1 \frac{\mathbf{G}^4 (r - 1)}{[1 + (r - 1)B]^5 \eta^5} W d\xi \tag{9.1.7}$$

where

$$\frac{dB}{d\xi} = W \tag{9.1.8}$$

Similarly, Eq. (9.1.6) is written as

$$\frac{d\eta}{d\xi} = \frac{L}{b_0} \frac{S_o + \frac{\mathbf{G}^2}{[1+(r-1)B]^3\eta^2}\left\{1 - \frac{\mathbf{G}^2}{[1+(r-1)B]^2\eta^3}\right\}\frac{(r-1)b_0 W}{L}}{1 - \frac{\mathbf{G}^2}{[1+(r-1)B]^2\eta^3}} \tag{9.1.9}$$

The design variable $B(\xi)$ has to satisfy the following conditions:

$$B(0) = 0 \tag{9.1.10}$$

$$B(1) = 1 \tag{9.1.11}$$

Furthermore, the state variable η at the inlet of transition is governed by the uniform flow in the pretransitional channel. Considering viscous effects unimportant, the nondimensional inlet depth $\eta(0)$ should satisfy Eq. (3.1.15), which is written as

$$\mathbf{G} = -2.457\eta(0)\sqrt{\frac{\eta(0)S_o}{1+2\eta(0)}}\ln\left[\frac{1+2\eta(0)}{12\eta(0)}\frac{\varepsilon}{b_0}\right] \tag{9.1.12}$$

Equation (9.1.12) can be solved by trial and error.

In order to avoid separation in the initial portion of the transition, the slope of transition at the inlet should be equal to zero. In the nondimensional variables, this condition is written as

$$W(0) = 0 \tag{9.1.13}$$

Furthermore, for avoiding the eddy formation in the final portion of the transition, the slope of the transition at the outlet should be zero, i.e.,

$$W(1) = 0 \tag{9.1.14}$$

Thus, the design problem boils down to minimization of H_L as given by Eq. (9.1.7) subject to constraints of Eqs. (9.1.8) through (9.1.14). This problem is an optimal control problem (Sage and White 1977).

In order to incorporate the equality constraint of Eq. (9.1.11), the following penalty function P_1 is added to the objective function:

$$P_1 = R_{1k}[B(1) - 1]^2 \tag{9.1.15}$$

where R_{1k} is a positive number. The equality constraints of Eqs. (9.1.13) and (9.1.14) can be taken care of by adding the following penalty function P_2 in the integrand of the objective function:

$$P_2 = \frac{R_{1k}W^2}{1+W^2}\left[(1+\xi)^{-16} + \xi^{-16}\right] \tag{9.1.16}$$

where R_{2k} = a positive number. The term in the square brackets of Eq. (9.1.16) has a peculiar behavior. For $\xi = 0$ and 1, its value in unity, and for all other values in the range $0 < \xi < 1$, it is nearly zero. Thus, the penalty is operative only at inlet and outlet of the transition. Combining the objective function and the penalty terms, the function is obtained as

$$J = R_{1k}[B(1) - 1]^2 + \int_0^1 \left\{ \frac{\mathbf{G}^4 (r-1)}{[1 + (r-1) B]^5 \eta^5} W \right. \\ \left. + \frac{R_{1k} W^2}{1 + W^2} \left[(1+\xi)^{-16} + \xi^{-16} \right] \right\} d\xi \tag{9.1.17}$$

The design problem, thus, reduces to minimization of J subject to Eqs. (9.1.8), (9.1.9), and (9.1.10) and Eq. (9.1.12). Combining the integrand of Eq. (9.1.17) with Eqs. (9.1.8) and (9.1.9) through the adjoint variables, λ_1 and λ_2, Hamiltonian H is obtained as

$$H = \frac{\mathbf{G}^4 (r-1)}{[1 + (r-1) B]^5 \eta^5} W + \frac{R_{1k} W^2}{1 + W^2} \left[(1+\xi)^{-16} + \xi^{-16} \right] + \lambda_1 W \\ + \lambda_2 \frac{L}{b_0} \cdot \frac{S_o + \frac{\mathbf{G}^2}{[1+(r-1)B]^3 \eta^2} \left\{ 1 - \frac{\mathbf{G}^2}{[1+(r-1)B]^2 \eta^3} \right\} \frac{(r-1) b_0 W}{L}}{1 - \frac{\mathbf{G}^2}{[1+(r-1)B]^2 \eta^3}} \tag{9.1.18}$$

The bed-width profile can be determined by solving the following adjoint equations:

$$\frac{d\lambda_1}{d\xi} = -\frac{\partial H}{\partial B} \tag{9.1.19}$$

$$\frac{d\lambda_2}{d\xi} = -\frac{\partial H}{\partial \eta} \tag{9.1.20}$$

and the optimality condition

$$\frac{\partial H}{\partial W} = 0 \tag{9.1.21}$$

The boundary conditions needed for solution of Eqs. (9.1.19) and (9.1.20) are given by the following equations (called transversality conditions):

$$\lambda_1(1) = 2R_{1k} [B(1) - 1] \tag{9.1.22}$$

$$\lambda_2(1) = 0 \tag{9.1.23}$$

9.1.1 Numerical Algorithm

Consider $n+1$ equispaced points $\xi_0, \xi_1, \xi_2, \ldots, \xi_n$, with $\xi_0 = 0$ and $\xi_n = 1$ and assume a trial profile of $W(\xi)$. Knowing $B(0)$ and $\eta(0)$ from Eqs. (9.1.10) and (9.1.12), respectively, Eqs. (9.1.8) and (9.1.9) are solved for $B(\xi_i)$ and $\eta(\xi_i)$ using a fourth-order Runge-Kutta method. Knowing $\lambda_1(1)$ and $\lambda_2(1)$ from Eqs. (9.1.22) and (9.1.23), respectively, Eqs. (9.1.19) and (9.1.20) are solved by a fourth-order Runge-Kutta method proceeding backward in ξ.

Using steepest descent method (Sage and White 1977), the change in W, ΔW is found as

$$\Delta W(\xi_i) = -K\frac{\partial H}{\partial W}; \quad i = 1, 2, 3, \ldots n. \tag{9.1.24}$$

where K = a small positive number. Assuming a small value of K, ΔW is worked out from Eq. (9.1.24) for all values of ξ_i. Knowing $\Delta W(\xi_i)$, new values of $W(\xi_i)$ are then obtained by

$$W_{k+1}(\xi_i) = W_k(\xi_i) + \Delta W(\xi_i); \quad i = 1, 2, 3, \ldots n. \tag{9.1.25}$$

where subscript k stands for the iteration number. The procedure is repeated till

$$\left|\frac{W_{k+1}(\xi_i) - W_k(\xi_i)}{W_{k+1}(\xi_i)}\right| \le \theta; \quad i = 1, 2, 3, \ldots n. \tag{9.1.26}$$

where θ = a predetermined small number. Varying the parameters $\mathbf{G}$, r, L/b_0, S_o, and ε/b_0 in practical ranges, a large number of bed-width profiles are found out by using the numerical algorithm. These profiles are fitted to the following equation:

$$b = b_0 + (b_L - b_0)\left[2.52\left(\frac{L}{x} - 1\right)^{1.35} + 1\right]^{-0.775} \tag{9.1.27}$$

Equation (9.1.27), when compared with the computed bed-width profiles, indicates that the maximum departure is less than 1 %. As the channel bed slope remains S_o throughout, there is no variation in the channel bed profile from the existing one. That is, the equation of bed profile is

$$z = -S_o x \tag{9.1.28}$$

where $z = 0$ is taken as the bed level of channel inlet.

Using a similar procedure, Swamee and Basak (1992) have found that for an expansive transition from a rectangular section to trapezoidal section of side slope m_L, the optimum bed width varies according to Eq. (9.1.27), and the optimum side slope follows as

$$m = m_L \left(\frac{x}{L}\right)^{1.23} \tag{9.1.29}$$

However, variation in the transition bed level was considered as given by Eq. (9.1.28). That is, the channel bed remains uninterrupted by the transition.

On the other hand, using a similar procedure, Swamee and Basak (1993) obtained the following equation for the channel bed profile:

$$z = -L\left\{\left[\frac{S_o x}{L} - 1.12\frac{z_L}{L}\left(\frac{x}{L}\right)^{1.12}\right]^{-10} + \left[-\frac{z_L}{L} - S_o\left(1 - \frac{x}{L}\right)\right]^{-10}\right\}^{-0.1}; \quad z_L < 0 \tag{9.1.30}$$

$$z = L\left\{\left[\frac{S_o x}{L} + 1.36\frac{z_L}{L}\left(\frac{x}{L}\right)^{1.13}\right]^{-10} + \left[\frac{z_L}{L} + S_o\left(1 - \frac{x}{L}\right)\right]^{-10}\right\}^{-0.1}; \quad z_L > 0 \tag{9.1.31}$$

where z_L/L is given by the following implicit equation:

$$\frac{z_L}{L} = 0.488\mathrm{e}^{0.358 m_L}\left(\frac{b_0}{b_L}\right)^{0.0257}\left(S_o^2 - \frac{z_L^2}{L^2}\right)^{0.5} - 161\,\mathrm{e}^{0.236 m_L}\left(\frac{b_0}{b_L}\right)^{0.104}\left(S_o^2 - \frac{z_L^2}{L^2}\right) \tag{9.1.32}$$

9.2 Contraction Transitions

It is a common practice to adopt the elliptical quadrant for a bed-width profile given by

$$b = b_0 + (b_0 - b_L)\sqrt{\xi\,(2 - \xi)} \tag{9.2.1}$$

Varshney et al. (1988) recommended the following form of Hartley et al.'s equation for the design of a rectangular to rectangular contraction transition:

$$b = \left[b_0^{-1} + \left(b_L^{-1} - b_0^{-1}\right)\xi\right]^{-1} \tag{9.2.2}$$

The elementary head loss dh_L, incurred in the elementary length dx, is composed of the surface-resistance loss dh_f and the form loss dh_m. That is,

$$dh_L = dh_f + dh_m \tag{9.2.3}$$

Denoting the friction factor by *f*, the elementary surface-resistance loss is written as

$$dh_f = \frac{fQ^2P}{8gA^3}dx \tag{9.2.4}$$

Neglecting viscous effects while using Eq. (3.1.13), for a rectangular canal, Eq. (9.2.4) is written as

$$dh_f = \frac{Q^2(b+2y)}{6gb^3y^3}\ln\left[\frac{\varepsilon(b+2y)}{12by}\right]^{-2}dx \tag{9.2.5}$$

Simon (1976) gave a table of form loss coefficient for an abrupt contraction of a circular conduit. Using these values, the form loss h_m, is expressed as

$$dh_m = 0.5\left[1-\left(\frac{A_L}{A_0}\right)^{1.175}\right]\frac{Q^2}{2gA_0^2} \tag{9.2.6}$$

where A_0 and A_L = inlet and outlet areas, respectively. Though Eq. (9.2.6) is obtained for a closed conduit, the same is assumed to be valid for an open-channel contraction. Considering an elementary length of a contraction transition and using Eq. (9.2.6), the elementary form loss is obtained as

$$dh_m = -\frac{Q^2dA}{3.4gA^3} \tag{9.2.7}$$

where dA = elementary increase in flow area. For a rectangular open channel, Eq. (9.2.7) takes the form

$$dh_m = -\frac{0.294Q^2}{gb^3y^2}\frac{db}{dx}dx \tag{9.2.8}$$

where the elementary increase in the flow depth dy has been neglected. Using Eqs. (9.2.3), (9.2.5), and (9.2.8), the head loss in the transition was found as

$$h_L = \int_0^L\left[\frac{Q^2(b+2y)}{6gb^3y^3}\ln\left[\frac{\varepsilon(b+2y)}{12by}\right]^{-2} - \frac{0.294Q^2}{gb^3y^2}w\right]dx \tag{9.2.9}$$

where w is given by Eq. (9.1.5). The governing differential equation of the gradually varied flow is given by Eq. (9.1.6). The boundary conditions imposed on the design variables are

$$b(0) = b_0 \tag{9.2.10}$$

$$w(0) = 0 \tag{9.2.11}$$

$$b(L) = b_L \tag{9.2.12}$$

$$w(L) = 0 \tag{9.2.13}$$

The state variable y at the inlet section $y(0) =$ the normal depth upstream of the transition, which can be obtained by Eq. (3.1.15), is written as

$$Q = -2.457by(0)\sqrt{\frac{gby(0)S_o}{b+2y(0)}}\ \ln\left[\frac{b+2y(0)}{12by(0)}\varepsilon\right] \tag{9.2.14}$$

The transition design, thus, boils down to the problem of minimization of h_L, as given by Eq. (9.2.9), subjected to the differential constraint Eq. (9.1.6) and the equality constraints given by Eqs. (9.2.10)–(9.2.14). This is an optimal control problem (Sage and White 1977).

The problem formulated in this section is similar to one worked out in the Sect. 9.1 for expansion transition design. By adopting the same numerical algorithm for a given set of data in the practical range, a large number of optimal contraction transition profiles are obtained. These profiles are fitted to the following equation:

$$b = b_0 + (b_0 - b_L)\left[1.41\left(\frac{L}{x} - 1\right)^{1.23} + 1\right]^{-0.924} \tag{9.2.15}$$

Comparing Eq. (9.2.15), with the computed bed-width profiles, it has been found that that the maximum departure is about 1.5 %. See Swamee and Basak (1994).

Adopting a similar procedure, Swamee and Basak (1994) have found that for a contraction transition from a trapezoidal section of side slope m_0 to a rectangular section, the optimum bed width varies according to Eq. (9.2.15) and the optimum side slope follows as

$$m = m_0\left[1 - \left(\frac{x}{L}\right)^{1.52}\right] \tag{9.2.16}$$

However in this formulation it was considered that the channel bed remains uninterrupted by the transition. Furthermore considering the variation of bed elevation also, it has been found that the variations in b and m are given by Eqs. (9.2.15) and (9.2.16), respectively; the variation in bed level is given by

$$z = -L\left\{\left[\frac{S_o x}{L} - 1.17\frac{z_L}{L}\left(\frac{x}{L}\right)^{1.2}\right]^{-10} + \left[-\frac{z_L}{L} - S_o\left(1 - \frac{x}{L}\right)\right]^{-10}\right\}^{-0.1};\quad z_L < 0 \tag{9.2.17}$$

$$z = L\left\{\left[-\frac{S_o x}{L} + 1.45\frac{z_L}{L}\left(\frac{x}{L}\right)^{1.5}\right]^{-10} + \left[\frac{z_L}{L} + S_o\left(1 - \frac{x}{L}\right)\right]^{-10}\right\}^{-0.1}; \quad z_L > 0 \tag{9.2.18}$$

where z_L/L is given by the following implicit equation:

$$\frac{z_L}{L} = 0.6\,\mathrm{e}^{0.41m_0}\left(\frac{b_0}{b_L}\right)^{0.026}\left(S_o^2 + \frac{z_L^2}{L^2}\right)^{0.5} - 173\,\mathrm{e}^{0.34m_0}\left(\frac{b_0}{b_L}\right)^{0.193}\left(S_o^2 + \frac{z_L^2}{L^2}\right)^{0.5} \tag{9.2.19}$$

References

Hartley GE, Jain JP, Bhattacharya AP (1940) Report on the model experiments of fluming of bridges on Purva branch, Technical memorandum 9. United Provinces Irrigation Research Institute, Roorkee, India, pp 94–98

Henderson FM (1966) Open channel flow. The MacMillan Co., New York

Sage AP, White CC (1977) Optimal systems control, 2nd edn. Prentice-Hall, Inc., Englewood Cliffs, pp 310–315

Simon AL (1976) Practical hydraulics. Wiley, New York, pp 59–63

Swamee PK, Basak BC (1991) Design of rectangular open-channel transitions. J Irrig Drain Eng ASCE 117(6):827–838

Swamee PK, Basak BC (1992) Design of trapezoidal expansive transitions. J Irrig Drain Eng ASCE 118(1):61–73

Swamee PK, Basak BC (1993) Comprehensive open-channel expansion transition design. J Irrig Drain Eng ASCE 119(1):1–17

Swamee PK, Basak BC (1994) Design of open-channel contraction transitions. J Irrig Drain Eng ASCE 120(3):660–668

Varshney RS, Gupta SC, Gupta RL (1988) Theory and design of irrigation structures, vol 2. Nem Chand and Bros, Roorkee, pp 85–98

Chapter 10
Optimal Design of Transmission Canal

Abstract A transmission canal conveys water from the source to a distribution canal. Many times, the area to be irrigated lies very far from the source, requiring long transmission canals. Though there is no withdrawal from a transmission canal, it loses water on account of seepage and evaporation. Hence, it is not economical to continue the same section throughout the length of a long transmission canal. Instead, a transmission canal should be divided into subsections or reaches, and the cross section for each of the subsections must be designed separately. This would result in reduced cross sections in the subsequent reaches. The reduced cross section not only results in cost saving for earthwork, lining, and water lost but also requires less cost in land acquisition, construction of bridges, and cross-drainage works. This chapter addresses the problem of design of transmission canal. Using the least-cost section equations presented in Chap. 8 and applying grid search method, equations for computation of the optimal subsection length and corresponding cost of a transmission canal have been obtained. The optimal subsection length of the transmission canal is independent of the length of the transmission canal. The optimal design equations along with the tabulated section shape coefficients provide a convenient method for the optimal design of a transmission canal. The present method can be extended in developing equations for the optimal design of a transmission canal having unequal cost of transitions and unequal length of subsections. The suggested equations are applicable for all the regular canal shapes. The section shape coefficients to be used in designing a transmission canal have been obtained for triangular, rectangular, and trapezoidal canals. The method can be extended to find the coefficients in the optimal design equations for other shapes such as the circular section, parabolic section, rounded corner trapezoidal section, etc., if the corresponding seepage functions are developed. Direct optimization procedures may be adopted for the optimal design of irrigation canal sections and for the transmission canal, but they are of limited use and require considerable amount of programming and computation. On the other hand, using the optimal design equations along with the tabulated section shape coefficients, the optimal design variables of a canal can be obtained in single-step computations.

Keywords Optimal canal section • Transmission canal • Feeder canal • Link • Transition • Dynamic programming • Grid search • Minimum cost • Section shape coefficient • Explicit design equations

P.K. Swamee, B.R. Chahar, *Design of Canals*, Springer Transactions in Civil and Environmental Engineering, DOI 10.1007/978-81-322-2322-1_10

Water or any other liquid is required to be carried over long distances through canals. Like electric transmission lines transmitting electricity, these canals transmit water. There are no intermediate withdrawals in a water transmission canal. Thus, a transmission canal is a long canal that does not have any intermediate withdrawal. This chapter discusses the design aspects of water transmission canals. As there are considerable losses on account of seepage and evaporation, it is uneconomical to design such a canal considering uniform discharge throughout its length. Thus, a transmission canal is divided into various links which are designed for uniform but progressively reducing link discharges.

10.1 Analytical Considerations

Consider a canal of length L_c, which is divided into N links of lengths $x_1, x_2, x_3, \ldots,$ x_N. Hence, there are $N-1$ transitions in the transmission canal. See Fig. 10.1. The cost of a transmission canal consists of the cost of links and the cost of transitions. Letting the cost of each transition C_T be the same for all the transitions and assuming canal geometry and depth to the drainage layer remain constant throughout each link, the overall cost of the transmission canal C_o would be

$$C_o = \sum_{i=1}^{N} C_i x_i + (N-1)\, C_T \tag{10.1.1}$$

where the index i indicates a link.

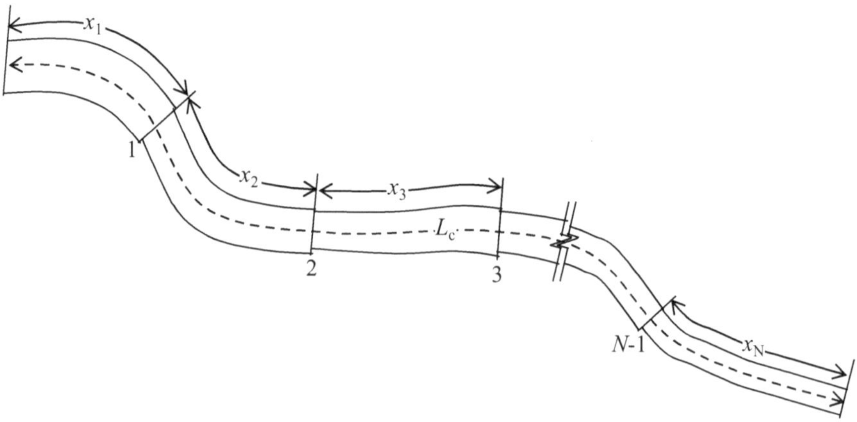

Fig. 10.1 Layout of a transmission canal

Using Eqs. (2.5.16), (3.1.15), and (10.1.1), the problem of the least-cost design of a transmission canal becomes (refer Swamee et al. 2002)

$$\text{Minimize} \quad C_o = \sum_{i=1}^{N} \left(c_e A_i + c_r A_i \overline{y_i} + c_L P_i + c_{ws} y_{n_i} F_{si} + c_{wE} T_i \right) x_i + (N-1) C_T \tag{10.1.2}$$

Subject to

$$2.457 A_i \sqrt{g R_i S_{oi}} \ln \left(\frac{\varepsilon}{12 R_i} + \frac{0.221 \nu}{R_i \sqrt{g R_i S_{oi}}} \right) + Q_i = 0, \quad \text{for all } i \tag{10.1.3}$$

$$Q_{i+1} - Q_i + \left(k y_{n_i} F_{si} + E T_i \right) x_i = 0, \quad \text{for all } i \tag{10.1.4}$$

$$\sum_{i=1}^{N} x_i - L_c = 0 \tag{10.1.5}$$

The first constraint, Eq. (10.1.3), imposes the condition of uniform flow in each link, while the second constraint, Eq. (10.1.4), satisfies the continuity in discharge from one link to the next link, and the third constraint, Eq. (10.1.5), is obvious. The optimization problem stated by Eqs. (10.1.2) through (10.1.5) is some sort of a dynamic programming problem. The problem is complicated due to an unknown number of links *N*, i.e., unknown number of constraints.

Assuming S_o, ε, ν, k, and E are constant for all the links and using a length scale λ_i as

$$\lambda_i = \left(\frac{Q_i^2}{g S_{oi}} \right)^{0.2} \tag{10.1.6}$$

and adopting $\lambda = \lambda_1$, the following nondimensional parameters are defined:

$$\lambda_{i*} = \lambda_i / \lambda \tag{10.1.7}$$

$$\varepsilon_* = \varepsilon / \lambda \tag{10.1.8}$$

$$\nu_* = \nu / \lambda \tag{10.1.9}$$

$$C_{o*} = C_e / \left(c_e \lambda^2 L_c \right) \tag{10.1.10}$$

$$c_{r*} = c_r\lambda / c_e \tag{10.1.11}$$

$$c_{L*} = c_L / (c_e\lambda) \tag{10.1.12}$$

$$c_{ws*} = c_{ws} / (c_e\lambda) \tag{10.1.13}$$

$$c_{wE*} = c_{wE} / (c_e\lambda) \tag{10.1.14}$$

$$C_{T*} = C_T / \left(c_e\lambda^2 L_c\right) \tag{10.1.15}$$

$$A_{i*} = A_i / \lambda_i^2 \tag{10.1.16}$$

$$\overline{y}_{ni*} = \overline{y}_{ni} / \lambda_i \tag{10.1.17}$$

$$P_{i*} = P_i / \lambda_i \tag{10.1.18}$$

$$T_{i*} = T_i / \lambda_i \tag{10.1.19}$$

$$R_{i*} = R_i / \lambda_i \tag{10.1.20}$$

$$x_{i*} = x_i / L_c \tag{10.1.21}$$

$$k_* = kL_c / \left(\lambda\sqrt{\lambda g S_o}\right) \tag{10.1.22}$$

$$E_* = E / k \tag{10.1.23}$$

Using Eqs. (10.1.7) through (10.1.23) in Eqs. (10.1.2) through (10.1.5), the optimization problem in nondimensional form is reduced to

$$\text{Minimize} \quad C_{o*} = \sum_{i=1}^{N} \left(A_{i*}\lambda_{i*}^2 + c_{r*}A_{i*}\overline{y_{i*}}\lambda_{i*}^3 + c_{L*}P_{i*}\lambda_{i*} + c_{ws*}y_{ni*}F_{si}\lambda_{i*} + c_{wE*}T_{i*}\lambda_{i*}\right)x_{i*} + (N-1)\,C_{T*} \tag{10.1.24}$$

$$\text{Subject to} \quad \Phi_i = -2.457A_{i*}\sqrt{gR_{i*}S_{oi}}\,\ln\left(\frac{\varepsilon_*}{12R_{i*}\lambda_{i*}} + \frac{0.221\nu_*}{R_{i*}\lambda_{i*}^{1.5}\sqrt{gR_{i*}S_{oi}}}\right), \quad \text{for all } i \tag{10.1.25}$$

$$\lambda_{i+1}^{2.5} - \lambda_i^{2.5} + k_* \left(y_{n*_i} F_{si} + E_* T_{i*}\right) x_{i*} = 0, \quad \text{for all } i \tag{10.1.26}$$

$$\sum_{i=1}^{N} x_{i*} - 1 = 0 \tag{10.1.27}$$

This dimensionless optimization problem is simplified to an optimization problem involving one variable *N*. This is achieved by providing the least-cost section for each link of the transmission canal and considering the length of each link to be equal.

Least-Cost Section Design Equations for Each Link

The cost of each section must be minimum to arrive at the least-cost transmission canal. The least-cost canal section for a particular section of the transmission canal could be obtained from Eq. (10.1.24), without the transition cost, subject to Eq. (10.1.25) and dropping the index *i* from them. The optimization problem in dimensionless form, thus, becomes

$$\text{Minimize} \quad C_* = A_* + c_{r*} A_* \overline{y_*} + c_{L*} P_* + c_{ws*} y_{n*} F_s + c_{wE*} T_* \tag{10.1.28}$$

$$\text{Subject to} \quad \Phi = -2.457 A_* \sqrt{R_*} \ln\left(\frac{\varepsilon_*}{12 R_*} + \frac{0.221 \nu_*}{R_* \sqrt{R_*}}\right) \tag{10.1.29}$$

The solution of this problem is given by Eqs. (8.1.8) through (8.1.11).

Optimal Cost and Length of Each Link

The use of Eqs. (8.1.8) through (8.1.11) for designing each section automatically satisfies the constraint given by Eq. (10.1.29). The optimization problem is further simplified by assuming the length of each subsection of the transmission canal to be the same, i.e., $x_i = x = L_c/N$ or $x_i^* = x^* = 1/N$ for all *i*. Thus, two constraints were eliminated and the problem becomes

$$\text{Minimize} \quad C_* = \frac{1}{N}\sum_{i=1}^{N} A_{i*} + c_{r*} A_{i*} \overline{y_{i*}} + c_{L*} P_{i*} + c_{ws*} y_{n_{i*}} F_{si} + c_{wE*} T_{i*} + (N-1) C_{T*} \tag{10.1.30}$$

$$\text{Subject to} \quad \lambda_{i+1}^{2.5} - \lambda_i^{2.5} + k_* \left(y_{ni*} F_{si} + E_* T_{i*}\right)/N = 0, \quad \text{for all } i \tag{10.1.31}$$

Once the optimal dimensions of a section are fixed using Eqs. (8.1.8) through (8.1.11), the discharge or λ for the next section becomes a function of *N* satisfying

Eq. (10.1.31). Now, the optimization problem stated by Eqs. (10.1.30) and (10.1.31) is left with finding the minimum of Eq. (10.1.30) for only one unknown, that is, N. By applying grid search (Fox 1971) on trapezoidal canal sections, a large number of optimal N were obtained for a number of input variables varying in the practical ranges ($10^{-5} \leq c_{T*} \leq \infty$ and $0 \leq k_* \leq 1.0$). Analysis of the optimal data so obtained resulted in the following equation for optimal number of links of the transmission canal for a trapezoidal channel:

$$N^* = L_c \sqrt{\frac{k + 0.15kL/d + 0.40E}{\sqrt{gLS_o}} \cdot \frac{2.18c_L + 4.10(1 + 0.21L/d)\, c_{ws} + 2.34c_{wE} + c_e L + 0.25c_r L^2}{15.65C_T}} \tag{10.1.32}$$

The value obtained from Eq. (10.1.32) is to be rounded to the nearest integer. If the optimal number of links happens to be zero or one, then no transition is required and assume $N^* = 1$. The optimal length of each link x^* of the transmission canal is given by

$$x^* = L_c/N^* \tag{10.1.33}$$

Further analysis of the optimal costs resulted in an empirical equation for the minimum cost of the transmission canal as

$$C_o^* = C^* L_c \left[1 + \left(\frac{0.433L_c}{L} \cdot \frac{k + 0.15kL/d + 0.48E}{\sqrt{gLS_0}} \right)^{1.28} \right.$$

$$\left. \frac{1.93c_L L + 3.45(1 + 0.16L/d)\, c_{ws} L + 0.97c_{wE} L + c_e L^2 + 0.19c_r L^3}{272C_T/L_c + 4.10c_L L + 7.202c_{ws} L + 1.56c_{wE} L + c_e L^2 + 0.17c_r L^3} \right]^{-1} \tag{10.1.34}$$

For a large depth of the drainage layer or water table and negligible evaporation loss, Eqs. (10.1.33) and (10.1.34) reduce to the following simplified equations:

$$x^* = \sqrt{\cdot \frac{15.65C_T \sqrt{gLS_o}}{(2.18c_L + 4.10c_{ws} + c_e L + 0.25c_r L^2)\, k}} \tag{10.1.35}$$

$$C_o^* = C^* L_c \left[1 + \left(\frac{0.433L_c k}{L\sqrt{gLS_0}} \right)^{1.28} \times \right.$$

$$\left. \frac{1.93c_L L + 3.45L_s c_w L + c_e L^2 + 0.19c_r L^3}{272C_T/L_c + 4.10c_L L + 7.202L_s c_w L + c_e L^2 + 0.17c_r L^3} \right]^{-1} \tag{10.1.36}$$

Similarly, other cases can be reduced from Eqs. (10.1.32) and (10.1.34).

Example 10.1 Design a 200-km-long trapezoidal transmission canal for carrying a discharge of 250 m^3/s on a longitudinal bed slope of 0.0001. The canal passes through a stratum of ordinary soil for which $c_e/c_r = 7$ m, $k = 5 \times 10^{-6}$ m/s, and depth of drainage layer = 7.5 m. Further, it is proposed to provide concrete lining with $c_\mathrm{L}/c_e = 12$ m. The climatic condition of the canal area is such that the maximum evaporation loss was estimated as 2.0×10^{-7} m/s (17.28 mm/day). Take rate of interest = 0.04 ₹/₹/year, $c_w/c_e = 1.267 \times 10^{-2}$, and $c_\mathrm{T}/c_e = 4 \times 10^5$.

Solution For the design, $g = 9.79$ m/s^2, $\nu = 1.007 \times 10^{-6}$ m^2/s (water at 20 °C), and $\varepsilon = 1$ mm (smooth-finished concrete lining) have been adopted. Using Eq. (5.1.9), $L = 23.950$ m.

Single-Link Design:

Using Eqs. (2.5.14) and (2.5.15), $c_{ws} = 3945.0c_w$ and $c_{w\mathrm{E}} = 157.8c_w$, respectively.

For a trapezoidal section, Table 8.1 gives section shape coefficients: for side slope $k_{m0} = 0.5774$, $k_{mL} = 14.2772$, $k_{mr} = 0.216$, $k_{ms1} = 23.494$, $k_{ms2} = 22.668$, $k_{md} = 0.124$, and $k_{mE} = 32.189$; for bed width $k_{b0} = 0.4341$, $k_{b\mathrm{L}} = 14.243$, $k_{br} = 0.348$, $k_{bs1} = 18.086$, $k_{bs2} = 14.416$, $k_{bd} = 0.130$, and $k_{b\mathrm{E}} = 0.288$; for normal depth $k_{y0} = 0.376$, $k_{y\mathrm{L}} = 14.227$, $k_{yr} = 0.348$, $k_{ys1} = 16.910$, $k_{ys2} = 14.885$, $k_{yd} = 0.112$, and $k_{y\mathrm{E}} = 4.036$; and for cost $k_{c0} = 0.245$, $k_{c\mathrm{L}} = 1.303$, $k_{cr} = 0.037$, $k_{cs} = 1.923$, $k_{cd} = 0.161$, and $k_{c\mathrm{E}} = 0.820$.

Using Eqs. (8.1.8), (8.1.9), and (8.1.10) with these coefficients: $m^* = 0.767$, $b^* = 17.065$ m, and $y_n{}^* = 6.264$ m. Further, Eq. (8.1.11) gives the canal cost per meter $C^* = 4{,}112.428c_e$. These dimensions yield $A = (b + my_n)y_n = 135.792$ m^2. Thus, $V = 250.0/135.792 = 1.841$ m/s, which is within the permissible limit. See Table 4.1.

Multiple-Link Design: Using Eq. (10.1.32) results: $N^* = 11.4$ rounded to 11; therefore, the optimal length of each link = 200/11 = 18.1818 km.

Optimal Section for the Second Link

Making use of design variables for the first link $m^* = 0.768$, $b^* = 17.055$ m, and $y_n{}^* = 6.220$ m and $d = 7.5$ m; in Eq. (2.4.18), the seepage function $F_s = 20.026$. Using Eq. (2.4.29), the water loss per unit length of the first link $q_w = 6.228 \times 10^{-4} + 5.322 \times 10^{-6}$ m^2/s. Thus, the total quantity of water lost in 18.1818 km length of the first link of the transmission canal = 11.420 m^3/s. Hence, the discharge available in the second link of the transmission canal = 250.0–11.420 = 238.58 m^3/s. Therefore, in the second link, $L = 23.524$ m. Using Eqs. (8.1.8), (8.1.9), (8.1.10), and (8.1.11) along with section shape coefficients from Table 8.1, $m^* = 0.764$, $b^* = 16.669$ m, $y_n{}^* = 6.137$ m, and $C^* = 4{,}013.673c_e$.

Likewise, the dimensions and cost per unit length of the canal for the successive subsections were calculated, and their variations along the length of the transition canal are shown in Table 10.1.

Table 10.1 Parameters of consequent links in a transmission canal

Link	Q	L	m	b	y	C	A	F_S	q_s (10^{-4})	q_E (10^{-6})	q_w
1	250.00	23.950	0.768	17.055	6.220	4,112.428	135.792	20.026	6.228	5.322	11.420
2	238.58	23.524	0.764	16.669	6.137	4,013.673	131.088	19.136	5.872	5.210	10.771
3	227.81	23.111	0.761	16.298	6.056	3,918.758	126.599	18.340	5.553	5.103	10.189
4	217.62	22.709	0.757	15.939	5.976	3,827.266	122.303	17.620	5.265	4.998	9.664
5	207.96	22.316	0.754	15.592	5.898	3,738.853	118.181	16.966	5.003	4.897	9.185
6	198.77	21.933	0.751	15.255	5.820	3,653.232	114.219	16.366	4.763	4.799	8.747
7	190.02	21.557	0.748	14.928	5.744	3,570.158	110.403	15.814	4.542	4.703	8.343
8	181.68	21.189	0.744	14.609	5.668	3,489.420	106.722	15.303	4.337	4.610	7.969
9	173.71	20.828	0.741	14.298	5.593	3,410.834	103.166	14.828	4.147	4.518	7.622
10	166.09	20.472	0.738	13.995	5.519	3,334.240	99.727	14.385	3.969	4.429	7.297
11	158.79	20.123	0.735	13.699	5.445	3,259.496	96.397	13.970	3.803	4.342	6.994

References

Fox RL (1971) Optimization methods for engineering design. Addison-Wesley Publishing Co., Reading, pp 43–44

Swamee PK, Mishra GC, Chahar BR (2002) Optimal design of a transmission canal. J Irrig Drain Eng ASCE 128(4):234–243

Chapter 11
Salient Features of Canal Route Alignment

Abstract An alignment of a canal route mainly depends on the topography. The total cost of a canal project depends upon the alignment. A canal has to be aligned in such a way that it covers the entire area proposed to be irrigated with the shortest possible length, and at the same time, its cost including the cost of cross-drainage works is a minimum. Several alignments between the source and the destination may be possible. Out of many alignments, few may not be feasible to construct due to construction-related problems. Deep cutting or high embankments are generally avoided by suitable detouring after comparing the overall costs of the alternative alignments. Land cost varies with land use pattern, resettlement and rehabilitation cost, environmental cost, and alignment of the canal; cost of canal falls/drops/cross-drainage works varies with type and size of structure. The maximization of economy is achievable by minimization of the total cost of canal route alignment considering all possible cost factors. This type of canal alignment problem is addressed in this chapter. Formulation of cost function comprising earthworks is difficult due to undulating terrain. In this chapter, the topography of a given area is expressed as a double Fourier series for overcoming this problem. A cost function is formulated along a canal route incorporating the depth of cutting to be extracted from the equation of the terrain and bed level of the canal. To arrive at a minimum cost route of a canal, several alignments can be evaluated by subdividing each alignment into segments. The chapter includes canal alignment algorithm, which can be followed to arrive at optimal canal route alignment.

Keywords Optimal canal alignment • Transition • Terrain • Topography • Route alignment • Fourier series • Minimum cost • Earthwork cost • Section cost • Balancing depth • Balancing length • Fall • Drop • Cross drainage • Cutting • Filling • Embankment • Cardano's method • Shoe function

The total cost of a canal project significantly depends upon its alignment. The alignment is the feasible path or route from a source location to the desired destination. A canal has to be aligned in such a way that it covers the entire area proposed to be irrigated with the shortest possible length, and at the same time, its cost including the cost of cross-drainage works is a minimum. A shorter length of canal ensures less loss of head due to friction and smaller loss of discharge due to

P.K. Swamee, B.R. Chahar, *Design of Canals*, Springer Transactions in Civil and Environmental Engineering, DOI 10.1007/978-81-322-2322-1_11

seepage and evaporation, so that additional areas can be brought under cultivation. Canal alignment may be contour canal, side slope canal, or ridge canal as per terrain of command area. A contour canal irrigates only one side of the canal, and it crosses a number of valleys. Thus, it involves different types of cross-drainage works like aqueducts, under tunnels, superpassages, etc. A side slope canal is aligned at right angles to the contours of a country. A watershed or ridge canal irrigates the areas on both sides. Cross-drainage works are completely eliminated in watershed and side slope canals. The main canal is carried on a contour alignment, till it reaches the top of a watershed to become a watershed canal.

A canal along the natural ground slope has the least excavation cost for canal section per unit length but as its length is more that increases the overall cost. On the other hand, the shortest route increases earthwork cost per unit length and cost of falls/drops and cross-drainage works, increasing the overall cost. Thus, there is a trade-off among all the alignments lying between these two extreme cases.

11.1 Unit Length Costs

11.1.1 Earthwork Cost

Equation (2.3.2) is modified as

$$C_e = c_e A_c + c_r A_c \overline{y_c} + c_f A_f + c_t \left[A_C - A_f \right] L_t \tag{11.1.1}$$

where A_c = flow area in cutting, c_f = cost per unit volume of earthwork in filling at ground level, A_f = flow area in filling, c_t = cost per unit volume of earthwork in haulage per unit distance at ground level, and L_t = length of haulage, i.e., the distance from working site to dump yard. The land cost C_{la} per unit length is expressed as

$$C_{la} = (c_{la} + c_{re} + c_{env})\, W \tag{11.1.2}$$

where c_{la} = cost of land per unit area, which depends upon land use like agricultural, residential, historic, forest, etc.; W = width of right of way of the canal section which depends upon the top width T and associated facilities; c_{re} = cost of rehabilitation per unit area, which depends upon the land use pattern and local laws; and c_{env} = cost of environment per unit area, which depends upon land use pattern and local environmental laws.

11.1.2 Unit Length Canal Section Cost

Combining Eqs. (2.2.1), (2.5.16), (11.1.1), and (11.1.2), the cost of canal per unit length C is obtained as

$$C_e = c_e A_c + c_r A_c \overline{y_c} + c_f A_f + c_t \left[A_c - A_f \right] L_t + (c_{la} + c_{re} + c_{env}) W$$
$$+ c_L P_l + c_{ws} y_n F_s + c_{wE} T \quad (11.1.3)$$

11.2 Balancing Depth Considerations

Earthwork in the form of cutting and/or filling along the canal alignment is required for providing canal flow area. For a fixed area section, the total canal cost per unit length is the sum of costs of earthworks in excavation and in embankments, which depend on the location of the water surface relative to the natural ground level. Canals are preferably aligned in partial cutting and partial filling to minimize the earthwork cost. The most economical depth or the balancing depth of cutting d_{cb} can be worked out by equating the quantity of material excavated to that required for constructing the banks.

For a unit length of canal in Fig. 11.1c,

$$(b + m_c d_{cb}) \, d_{cb} = 2 \left(b_w + m_f d_f \right) d_f \quad (11.2.1)$$

where $b =$ bed width of canal, $b_w =$ top width of banks, $m_c =$ canal side slope in cutting, $m_f =$ bank slope in filling, and $d_f =$ filling height in banks. From the terrain equation and canal bed line equation, the depth of cut d_c or the depth of fill $-d_c$ can be determined. This d_c must be close to d_{cb} for economy; hence, for a fixed area section, the cost function becomes $K|d_{cb} - d_c|$, where $K =$ constant, which is a function of shape of cross-sectional area and soil type (i.e., c_e, c_r, s_c, and s_f).

Between the source and destination, there may be several possible routes for the canal alignment. For example, Fig. 11.2 shows three routes, SAD, SBD, and SCD, between the source S (15 km, 0 km, 100 m) and the destination D (60 km, 100 km, 50 m). Deep cutting or high embankments can be avoided by suitable detouring after comparing the overall costs of the alternative alignments. A canal alignment can be divided into various reaches say N_d. Each reach can be between two cross-drainage/drop structures with constant Q and S_o, and thus, it will require $N_d - 1$ cross-drainage/drop structures. Therefore, the total cost of a canal alignment C_t is

$$C_t = \sum_{i=1}^{N_d} \sum_{j=1}^{N_s} K_{ij} \left| d_{cbij} - d_{cij} \right| x_{ij} + \sum_{i=1}^{N_d - 1} C_{di} \quad (11.2.2)$$

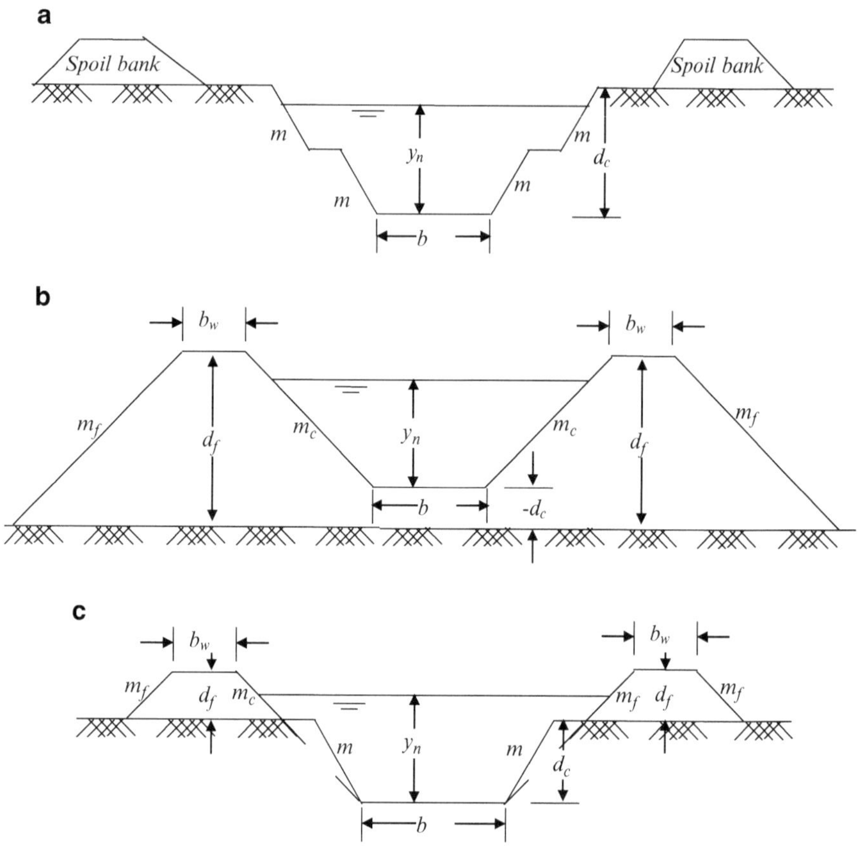

Fig. 11.1 Typical canal sections. (**a**) Canal in full cutting. (**b**) Canal in full filling. (**c**) Canal in partial cutting and partial filling

where N_d = number of reaches/divisions, N_s = number of segments in ith reach, x_{ij} = length of the jth segment in ith reach, and C_{di} = cost of ith cross-drainage/drop structure. The cost of a drop structure will depend on the depth of fill upstream of the structure and depth of cutting downstream of the structure. Considering various alignments, the minimum cost alignment can be finalized.

Example 11.1 A trapezoidal canal section is to be constructed, having 5.0 m bottom width, 2.5 m normal depth, 1:1.25 (V:H) side slope, and 0.5 m free board. Determine the balancing cutting depth to have 2.0 m top width and 1:2 slope for compacted banks on both sides of the canal.

Solution The filling height of banks is such that $d_c + d_f = y_n + F = 2.5 + 0.5 = 3.0$. Using Eq. (11.2.1) with $b = 5.0$ m, $m_c = 1.25$, $b_w = 2.0$, and $m_f = 2$, $(5 + 1.25d_{cb})\, d_{cb} = 2 \times (2 + 2 \times (3 - d_{cb}))\,(3 - d_{cb})$ which yields the balancing depth $d_{cb} = 1.694$ m and hence height of banks $d_f = 1.306$ m.

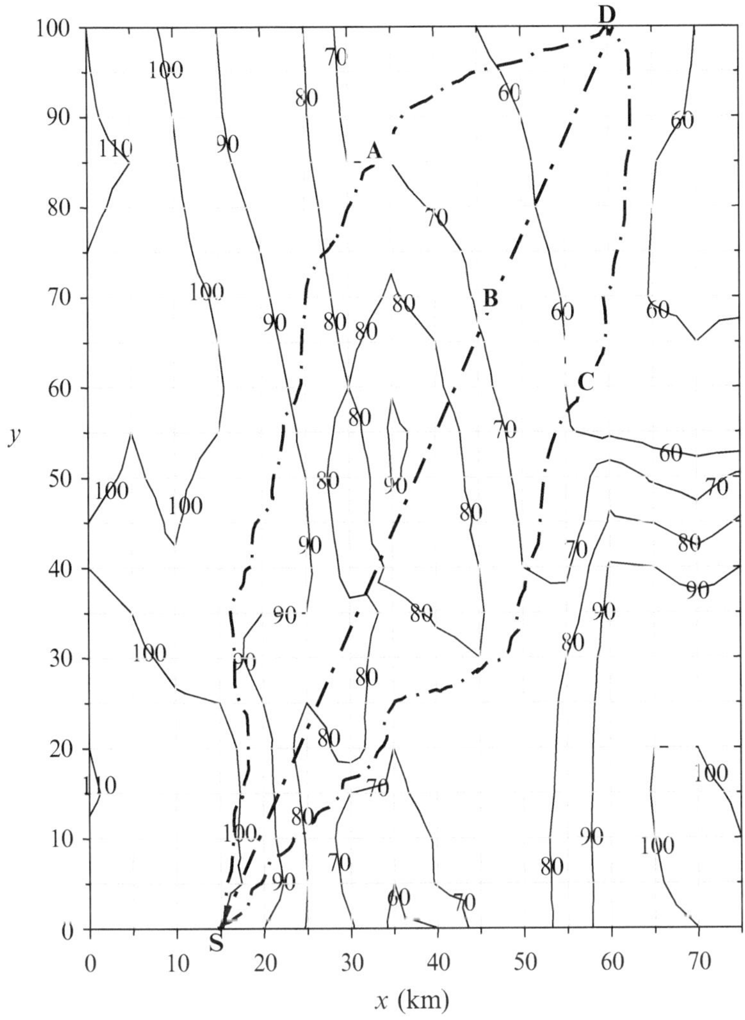

Fig. 11.2 Canal route alignments

11.3 Balancing Length Considerations

To reduce earthwork cost and other associated implications, the earthwork should be balanced in excavation and filling for a feasible haulage length along the selected route of the canal. The total canal length can be divided into segments to balance the earthwork within that segment. To find out the initial cutting height, the cutting and filling volumes have been balanced for that sub-reach. It has been assumed that the whole cutting volume can be used for filling purpose with desired compaction without any wastage. The length l of the canal segment/reach has been divided into three parts, namely, l_1, l_2, and l_3 as shown in Fig. 11.3 (Basu 2013).

The first part l_1 is the length of the sub-reach where the canal section is in full cutting (see Fig. 11.4) with an inside *berm* of width b_{w1}, cutting depth d_c, and side slope as m_c. At this part, d_c is always greater than $y_n + F_B$.

The second subdivision l_2 is the length where canal runs in partial cutting and filling (Fig. 11.5). At this part, d_c is always less than $y_n + F_B$.

The third subdivision l_3 is the length where the canal runs in complete filling (Fig. 11.6) with filling embankment top width as b_w, height as d_f, and side slope as m_f. At this part, d_c is always negative, and d_f is greater than $y_n + F_B$.

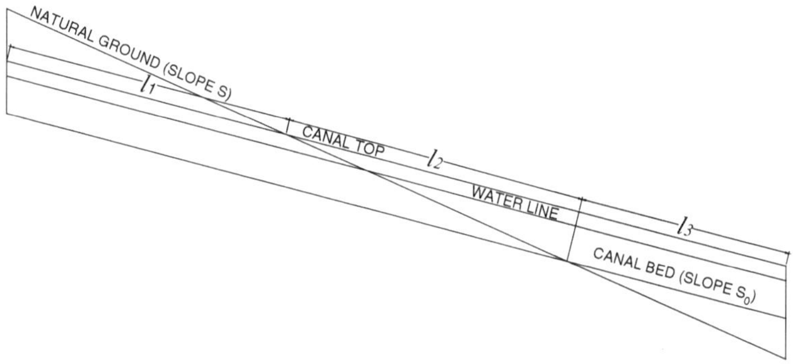

Fig. 11.3 Longitudinal section of canal

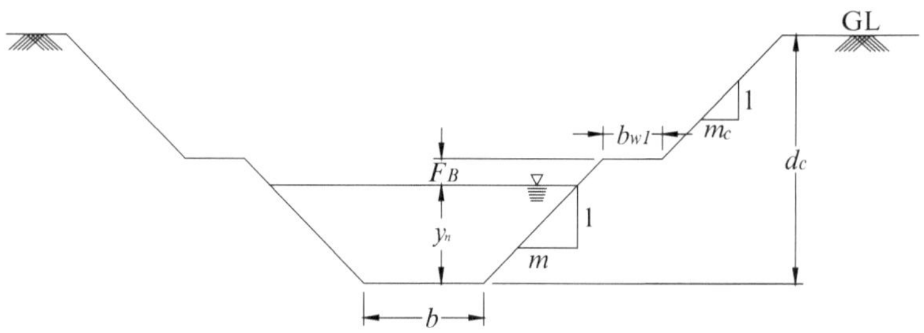

Fig. 11.4 Trapezoidal canal section in cutting

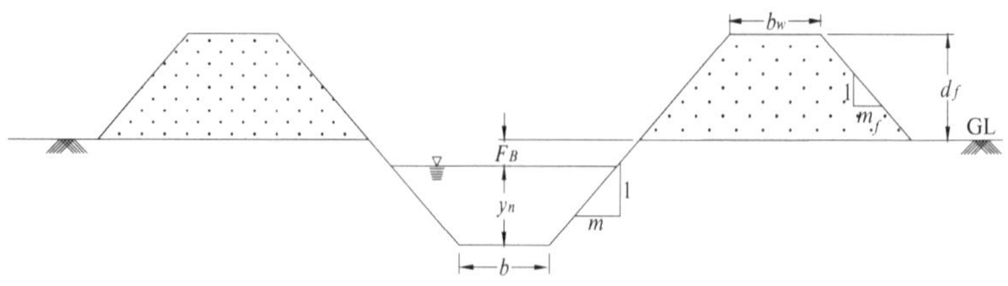

Fig. 11.5 Trapezoidal canal section in cutting and filling

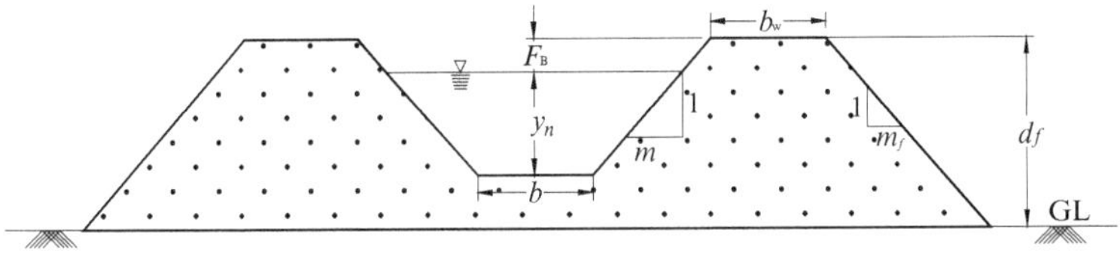

Fig. 11.6 Trapezoidal canal section in filling

l_1, l_2, and l_3 of a sub-reach are related to depth of cutting, height of filling, canal bed slope S_o, and natural ground slope S_g as

$$l_1 = (d_c - y_{FB}) / S_d \tag{11.3.1}$$

$$l_2 = y_{FB} / S_d \tag{11.3.2}$$

$$l_3 = \left(d_f - y_{FB}\right) / S_d \tag{11.3.3}$$

$$d_c = l S_d + y_{FB} - d_f \tag{11.3.4}$$

where $S_d = S_g - S_0$ is slope difference and $y_{FB} = y_n + F_B$ is water depth including free board.

The total volume of cutting V_c for a trapezoidal canal section in a sub-reach is expressed as

$$V_c = (b + m y_{FB}) y_{FB} (l_1 + 0.5 l_2) \quad + 0.5 \{b + 2b_{w1} + 2m y_{FB} + m_c (d_c - y_{FB})\} (d_c - y_{FB}) l_1 \tag{11.3.5}$$

The total volume of filling V_f is expressed as

$$V_f = \left(2b_w + m y_{FB} + m_f y_{FB}\right) y_{FB} (l_3 + 0.5 l_2) + 0.5 \ \{b + 2b_w + 2m y_{FB} + 2m_f y_{FB} + m_f \left(d_f - y_{FB}\right)\} \left(d_f - y_{FB}\right) l_3 \tag{11.3.6}$$

Mainly, the initial cutting depth governs the cost of fall as well as cost of earthwork within a sub-reach as it is the controlling parameter in deciding the type of fall and the volume of balancing earthwork. The initial cutting depth and corresponding filling height can be determined by balancing the earthwork volume within a sub-reach. Equating Eqs. (11.3.5) and (11.3.6) and solving for balancing earthwork, we get the following cubic equation in d_f:

$$\left(m_c + m_f\right) h_f^3 + P_{T1} h_f^2 + P_{T2} h_f - P_{T3} = 0 \tag{11.3.7}$$

where

$$P_{T1} = 2b_W - 2b_{W1} - m_f y_{FB} - 3m_c l S_d \tag{11.3.8}$$

$$P_{T2} = 3m_c l^2 S_d^2 + 2\left(b + 2b_{w1} + 2my_{FB}\right) l S_d + m_f y_{FB}^2 \tag{11.3.9}$$

$$P_{T3} = m_c l^3 S_d^3 + \left(b + 2b_{W1} + 2my_{FB}\right) l^2 S_d^2 + 2\left(b + my_{FB}\right) y_{FB} l S_d \tag{11.3.10}$$

The feasible root of Eq. (11.3.7) providing the optimal height of filling d_{fo} can be computed using Cardano's method as described in Appendix 3, and then the corresponding optimal depth of cutting d_{co} is obtained using Eq. (11.3.4).

Example 11.2 A trapezoidal canal section has 5.0 m bottom width, 2.5 m normal depth, 1:1.25 (V:H) side slope, 1:1,000 bed slope, and 0.5 m free board. Calculate the optimal cutting and filling heights for 1,000 m length of the canal if berm width = 1.5 m and cutting slope = 1:1.5 in cutting part and top width of banks = 2.0 m and embankment slope = 1:2 in filling part. Assume natural ground slope = 1:100.

Solution Given data: $b = 5.0$ m, $y_n = 2.5$ m, $m = 1.25$, $F_B = 0.5$ m, $S_0 = 1/1{,}000$, $l = 1{,}000$ m, $m_c = 1.5$, $m_f = 2$, $b_w = 2.0$ m, $b_{w1} = 1.5$ m, and $S_g = 1/100$. Using these data in Eqs. (11.3.8), (11.3.9), and (11.3.10), $P_{T1} = -45.5$, $P_{T2} = 661.5$, and $P_{T3} = 2821.5$. Following Appendix 3 and using Eqs. (A3.1.3) and (A3.1.4), $A = 132.667$ and $B = -149.884$, so $\frac{B^2}{4} + \frac{A^3}{27} = 92097.47 > 0$ indicates one real root of the equation given by $\left(-\frac{B}{2} + \sqrt{\frac{B^2}{4} + \frac{A^3}{27}}\right)^{\frac{1}{3}} + \left(-\frac{B}{2} - \sqrt{\frac{B^2}{4} + \frac{A^3}{27}}\right)^{\frac{1}{3}} - \frac{P_{T1}}{3(m_c + m_f)}$. Hence, the optimal filling height $h_{fo} = 5.45$ m and the optimal cutting height $h_{co} = 6.55$ m.

11.4 Canal Alignment: Costs

When a mild slope canal passes through a steep slope terrain, the canal bed level becomes higher than the existing natural ground level and involves heavy filling. To avoid this fall, structures are envisaged whenever the elevation of the bed is higher than the existing natural ground level. Also, the canal cross-sectional dimensions may be revised after the fall structures based on the revised flow considering seepage and evaporation loss of the previous sub-reaches. The type fall structure is the function of the discharge and the height of the fall h. The height of fall should be such that the after fall, the full supply level should reach in normal by 500 m in downstream of the canal.

As a general guide, for deciding upon the type of the cross-drainage work, important considerations are full supply level and functions of canal vis-à-vis high flood level of the drainage channel, topography of terrain, regime of the stream, foundation strata, dewatering requirements, ratio of design flood to be provided for the drainage channel to the discharge in the canal, and envisaged head loss. When the bed level of the river is lower than the bed level of the canal, then aqueduct or siphon aqueduct is required else superpassage is provided. When the water levels of the river and the canal are the same, then level crossing is preferred. Similarly, a small culvert or a bridge depending upon the canal size and flow through the canal is required when a canal crosses a road or railway line.

Several alignments are possible between the source and the destination. As shown in Fig. 11.7, three alternate alignments are shown and they have their own merits and demerits. Alternate 3 is passing through forest land, and acquisition of which may be very difficult and expensive. Alternate 2 is the shortest route, but it passes through a forest land and a historical structure. The demolishing of the historical structure may not be allowed, and the acquisition of the forest land may be problematic. Alternate 1 has one fall, one aqueduct, and one superpassage as shown in Fig. 11.8, and it is the longest route as the canal is a contour canal. Thus, the cost of three alternatives will be different from each other, and hence, the alignment of a canal route plays an important role in the overall cost of the canal project (Basu and Chahar 2010).

11.5 Terrain Representation by Fourier Series

The cost function is a function of the domain topography. The topography is a function which represents elevation as a function of plan coordinates. The elevation can be expressed as double sine series of plan area (x, y) which is a continuous function of plan coordinates. Considering distances in kilometers, the coordinates of the area under consideration are A(−150.049, 333.580), B(157.088, 104.244), C(76.575, −3.683), and D(−231.013, 225.803). Plotting the coordinates, the resulting area is depicted in Fig. 11.9. It can be seen that the area under consideration is an inclined rectangle in shape with the angle of inclination being θ. Changing the origin of the coordinates to (x_D, y_D), the new coordinates (x', y') are given by

$$x' = x - x_D \tag{11.5.1}$$

$$y' = y - y_D \tag{11.5.2}$$

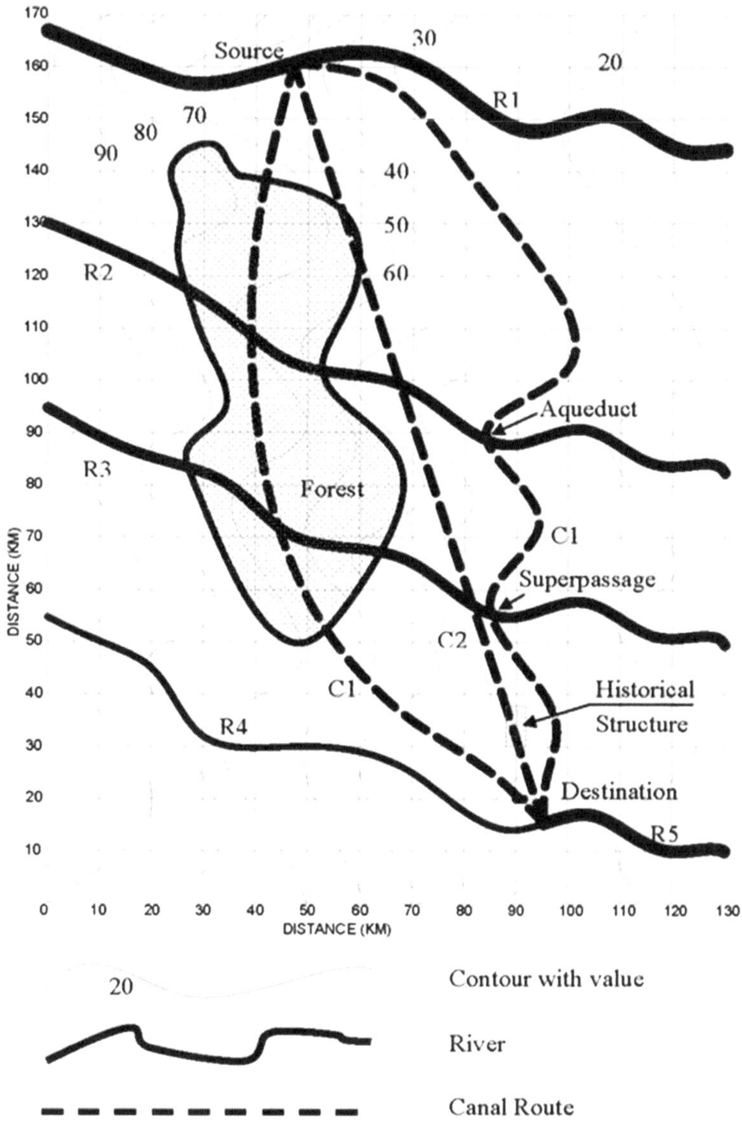

Fig. 11.7 Alternate routes for alignment of a canal

Rotating these coordinates counterclockwise by an angle θ, the new coordinates (u, v) are given by

$$u = x' \cos\,\theta + y' \sin\,\theta \tag{11.5.3}$$

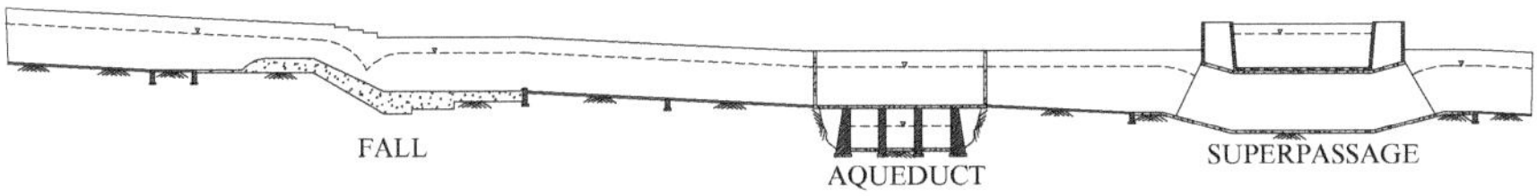

Fig. 11.8 Longitudinal section of alternative 1 with possible location of fall, aqueduct, and superpassage

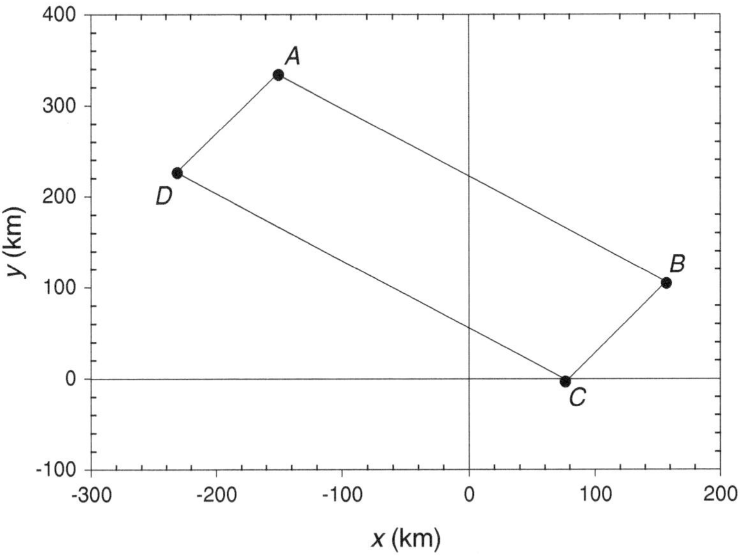

Fig. 11.9 The area of interest in the original coordinates (x, y)

$$v = -x'\sin\,\theta + y'\cos\,\theta \tag{11.5.4}$$

Combining Eqs. (11.5.1), (11.5.2), (11.5.3), and (11.5.4), the following equations of coordination transformation are obtained:

$$u = (x - x_D)\cos\,\theta + (y - y_D)\sin\,\theta \tag{11.5.5}$$

$$v = -(x - x_D)\sin\,\theta + (y - y_D)\cos\,\theta \tag{11.5.6}$$

The inverse transformation is

$$x = x_D + u\,\cos\theta - v\,\sin\theta \tag{11.5.7}$$

$$y = y_D + u\,\sin\,\theta + v\,\cos\,\theta \tag{11.5.8}$$

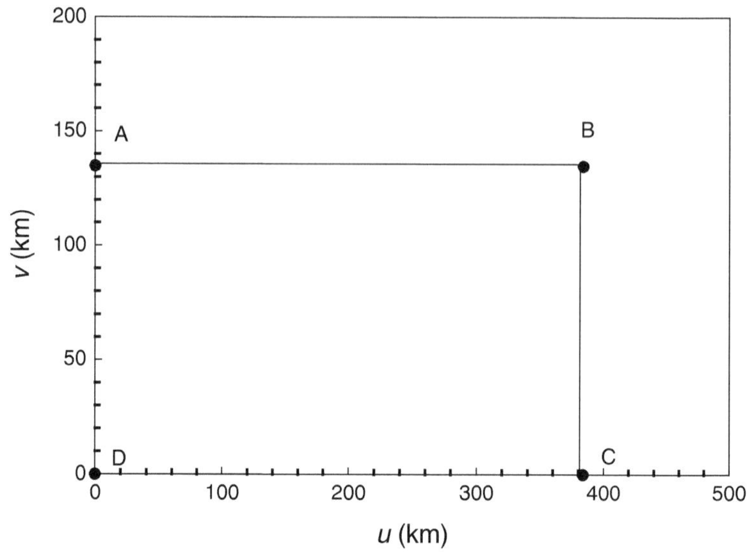

Fig. 11.10 The area as seen in the modified coordinates (u, v)

Using Eqs. (11.5.5) and (11.5.6), the points A, B, C, and D are shifted to another location, forming a rectangle with the following coordinates: A(0, 134.799), B(383.764,134.799), C(383.764, 0), and D(0,0). See Fig. 11.10. Thus, the rectangle is having the length $L = 383.764$ m and width $B = 134.799$ m.

The data constitutes 15,062 elevation points in the area of interest. The coordinates of the data vary in an arbitrary manner. Considering a mesh of $(n+1) \times (n+1)$ grid points with $\Delta u = L/n$ and $\Delta v = B/n$, elevation z_{ij} at regular grid point $P(i\Delta u, j\Delta v)$ is obtained by using

$$z_{ij} = \frac{z_p/r_p + z_q/r_q + z_r/r_r + z_s/r_s}{1/r_p + 1/r_q + 1/r_r + 1/r_s} \tag{11.5.9}$$

where r_p, r_q, r_r, and r_s = four points closest to the point P.

The elevation $z(u, v)$ of the topography can be expressed in terms of double Fourier series consisting of sine and cosine terms. Thus, there will be four series consisting of sine-sine, sine-cosine, cosine-sine, and cosine-cosine terms. On the other hand, if the topography function to be expanded vanishes on all the four sides of the rectangular area, the function can be expanded by sine-sine series only. Besides eliminating the Gibb's oscillations near the edges, such an expansion saves considerable amount of computation work. For this purpose, the elevation difference function $z_D(u, v)$ is defined as

$$z_D(u, v) = z(u, v) - \varsigma(u, v) \tag{11.5.10}$$

where $\varsigma(u, v)$ = a function that matches with $z(u, v)$ at all the four edges and vertices and varies smoothly inside the area. This function may be called periphery-matching function. Such a function can be written as (Swamee and Chahar 2013)

$$\varsigma\,(u,v) = N_B z\,(u,0) + N_T z\,(u,B) + N_L z\,(0,v) + N_R z\,(L,v) \qquad (11.5.11)$$

where N_B = bottom side function that is unity at the bottom and on all other three sides it vanishes, and N_T, N_L, and N_R =, respectively, the top-side, left-side, and right-side functions having similar properties. The functions N_B, N_T, N_L, and N_R are not unique; however, the following functional forms have been conceived:

$$N_B = \left\{1 + \left(\eta^2 + p\right)\left[\frac{1}{\xi^2 + p} + \frac{1}{(1-\xi)^2 + p} + \frac{1}{(1-\eta)^2 + p}\right]\right\}^{-1} \qquad (11.5.12)$$

$$N_T = \left\{1 + \left[(1-\eta)^2 + p\right]\left[\frac{1}{\xi^2 + p} + \frac{1}{(1-\xi)^2 + p} + \frac{1}{\eta^2 + p}\right]\right\}^{-1} \qquad (11.5.13)$$

$$N_L = \left\{1 + \left(\xi^2 + p\right)\left[\frac{1}{(1-\xi)^2 + p} + \frac{1}{\eta^2 + p} + \frac{1}{(1-\eta)^2 + p}\right]\right\}^{-1} \qquad (11.5.14)$$

$$N_R = \left\{1 + \left[(1-\xi)^2 + p\right]\left[\frac{1}{\xi^2 + p} + \frac{1}{\eta^2 + p} + \frac{1}{(1-\eta)^2 + p}\right]\right\}^{-1} \qquad (11.5.15)$$

where $\xi = u/L$, $\eta = v/B$, and $p = \xi\eta\,(1-\xi)\,(1-\eta)$. Applying Eq. (11.5.11), using Eqs. (11.5.12), (11.5.13), (11.5.14), and (11.5.15), it is found that as the corner point (0, 0) is included both in N_B and N_L, and each function is unity, $\zeta(0, 0) = 2z(0, 0)$. Similarly, $\zeta(0, L) = 2z(0, L)$, $\zeta(B, 0) = 2z(B, 0)$, and $\zeta(B, L) = 2z(B, L)$. Thus, the function $\zeta(u, v)$ does not match at the corner points. For matching at the corner points, for these points, N_B, N_T, N_L, and N_R are modified to 0.5. Figure 11.11 depicts a plot of modified function N_B. It can be seen that as the function resembles a shoe, it may be called a *shoe function*.

The periphery-matching function $\zeta(u, v)$ is shown plotted in Fig. 11.12. It is found that the elevation of the rectangular boundary of the topography and this function are the same. Figure 11.13 depicts the subtracted function z_D. A perusal of Fig. 11.13 reveals that the elevations of the rectangular boundary are zero. Thus, z_D

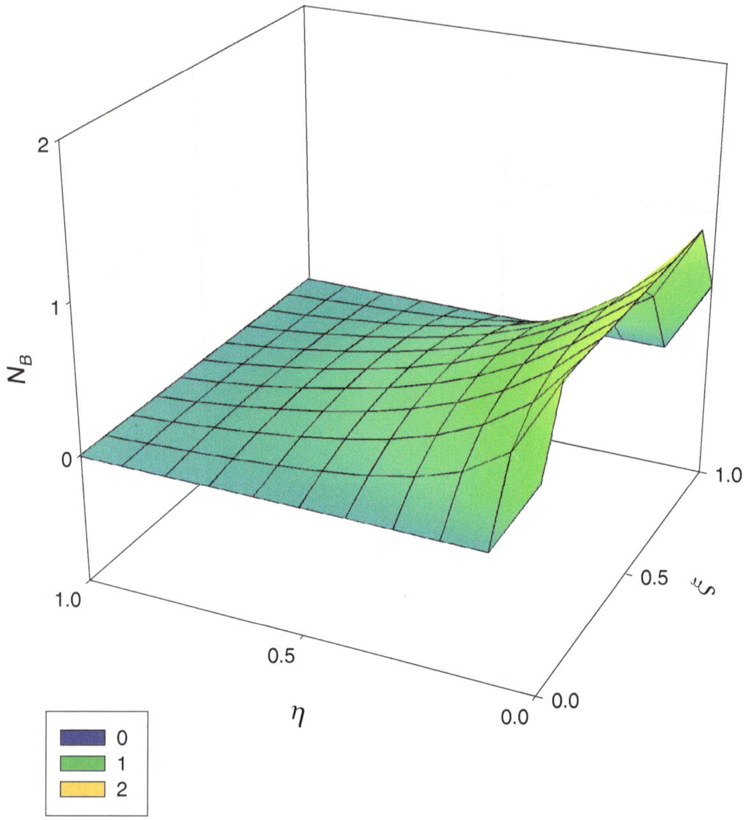

Fig. 11.11 The shoe function N_B

can be expressed as a double Fourier series of sine terms only. That is, the terrain elevation $z(u, v)$ is expressed as the following double Fourier series containing product of two sine factors:

$$z(u, v) = \zeta(u, v) + \sum_{l=1}^{l_A} \sum_{k=1}^{k_A} b_{Akl} \sin\left(\frac{k\pi u}{L}\right) \sin\left(\frac{l\pi v}{B}\right) \tag{11.5.16}$$

where k_A and l_A = cutoff frequencies of k and l, respectively, and b_{Akl} = Fourier coefficient for k and l. The Fourier coefficient b_{Akl} is given by

$$b_{Akl} = \frac{4}{LB} \int_0^B \int_0^L z_D(u, v) \sin\left(\frac{k\pi u}{L}\right) \sin\left(\frac{l\pi v}{B}\right) du\,dv \tag{11.5.17}$$

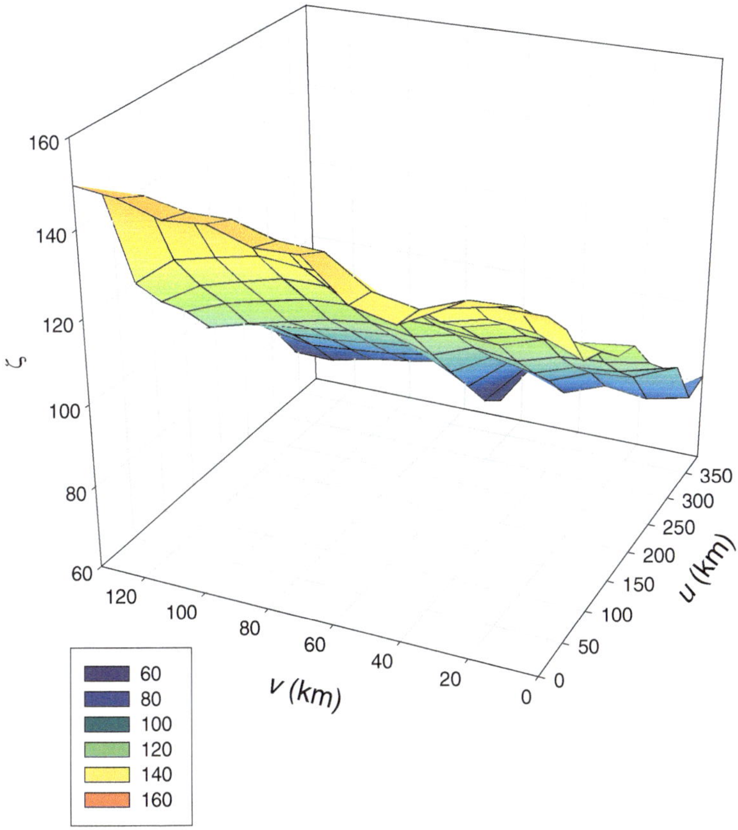

Fig. 11.12 The periphery-matching function

where $z_D(u, v)$ is given by Eqs. (11.5.10), (11.5.11), (11.5.12), (11.5.13), (11.5.14), and (11.5.15). Eq. (11.5.17) can be converted to the following form involving summation:

$$b_{Akl} = \frac{4}{LB}\sum_{j=0}^{n-1}\sum_{i=0}^{m-1} z_D\left(u_i, v_j\right)\sin\left(\frac{k\pi u_i}{L}\right)\sin\left(\frac{l\pi v_j}{B}\right) \tag{11.5.18}$$

where i and j = indices for u and v, respectively. Using Eq. (11.5.18), the Fourier coefficients b_{Akl} can be obtained. Using Eqs. (11.5.5), (11.5.6), and Eq. (11.5.11) through Eq. (11.5.15), Eq. (11.5.16) is expressed as a function of x and y. Such a function is fairly complicated.

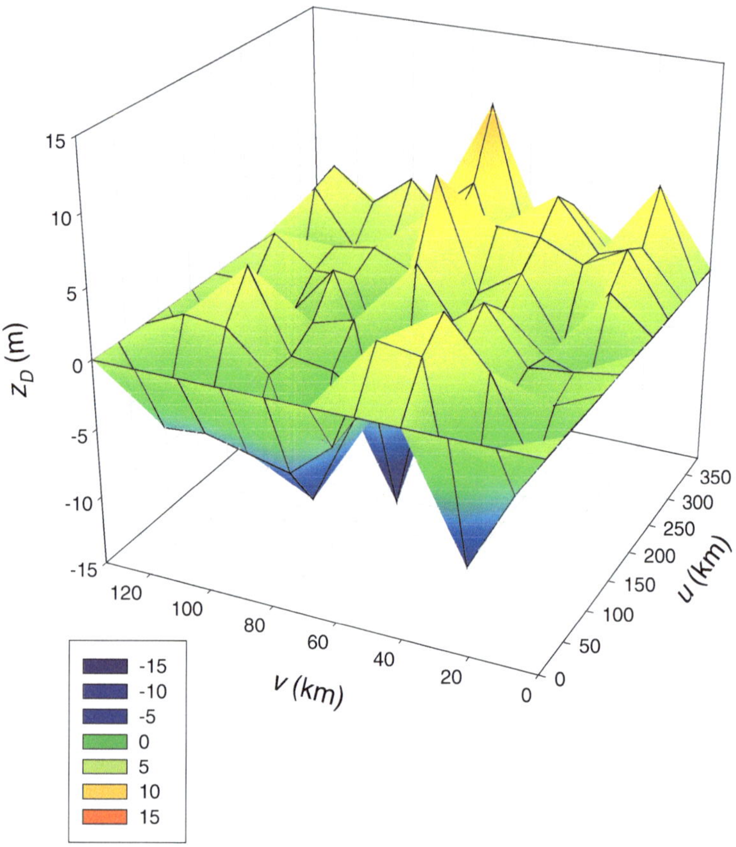

Fig. 11.13 The difference function

11.5.1 Canal Alignment Algorithm

Canal alignment is important for irrigation canals. Several alignments may be possible between the source and destination of a canal, but the minimum cost alignment is an obvious choice. Here, optimization of canal route alignment between the head works (source S) and the watershed (destination D) is discussed. The destination point D (Fig. 11.7) can be varied on the watershed to derive the optimum solution. The topography of a given area can be expressed as a double Fourier series of sine terms. In the method applied for this optimization, the topography of a given area was expressed as a double Fourier series of sine terms with the Fourier series representation being used to find the depth of cut or fill along a route of a canal alignment. Several alignments are then evaluated to get the minimum cost route for the canal. Overall, the implementation of the method involves the following procedure (Swamee and Chahar 2013):

Step 1 Mark a rectangular area of interest *ABCD* on the map. Prepare elevation data (x_i, y_i, z_i) for $i = 1 - N$ on this area.
Step 2 Using Eqs. (11.5.5) and (11.5.6), translate and orient the area to the new location and find B and L.
Step 3 Using Eq. (11.5.9), find elevations on rectangular grid points (u_i, v_i).
Step 4 Using Eqs. (11.5.12), (11.5.13), (11.5.14), and (11.5.15), find functions N_B, N_T, N_L, and N_R on all the grid points. Modify these values at the corner points by multiplying by 0.5.
Step 5 Using Eq. (11.5.11), find the periphery-matching function for all grid points.
Step 6 Find z_D using Equation (11.5.10) for all grid points.
Step 7 Using Eq. (11.5.18), find Fourier coefficients $b_{A\,kl}$ for, say, $k = 1$–10 and $l = 1$–10.
Step 8 Using Eq. (11.5.16), find the equation of topography.
Step 9 Adopt a trial canal alignment and divide it into several reaches. Adopting bed slopes in these reaches, find the bed line.
Step 10 The difference between elevations of source and destination yields the total drop required, while the bed slope multiplied by alignment length giving total drop in bed level would be acquired. The additional drop is attained by providing suitable fall structures.
Step 11 Design canal sections for all the reaches using a standard method of optimal canal section design. This obtains bed widths and side slopes in all the reaches.
Step 12 For each reach, adopting balancing depth and balancing length concepts, compute canal cost using Eq. (11.1.3). Repeat this for all the reaches.
Step 13 Find the cost of alignment by summing cost of all reaches and adding cost of fall structures and cross-drainage works required in the alignment.
Step 14 Choose another alignment and go to step 9.
Step 15 Select minimum cost alignment out of several alignments generated by steps 9–14.

References

Basu S (2013) Optimal alignment of a canal route. Ph.D. thesis, Indian Institute of Technology Delhi, New Delhi

Basu S, Chahar BR (2010) Cost function of canal route alignment. In: Conference proceedings, EWRI of ASCE '3rd international perspective on current and future state of water resources and environment, Chennai, India, 5–7 January 2010

Swamee PK, Chahar BR (2013) Optimal alignment of a canal route. Proc Inst Civil Eng Water Manag 166(WM8):422–431

Appendices

Appendix 1: Lambert's W Function

In 1758, Johann Heinrich Lambert (1728–1777) first considered an equation that led Leonhard Euler (1707–1783) in 1783 to derive a function W(x) as a solution of the equation

$$We^{W} = x \tag{A1.1}$$

Lambert's W function is shown plotted in Fig. A1.1. A perusal of Fig. A1.1 reveals that the principal branch of Lambert's W function, denoted by W_0, starts from the point (-1/e, -1) and is monotonically increasing function of x. There is a subsidiary branch denoted by W_{-1} that also starts from (-1/e, -1) and is asymptotic to the w-axis.

Lambert's W function finds applications in various branches of mathematics, statistics, physics, and engineering. On account of its importance, the Lambert's W function is included in many software like Maple and Mathematica.

Solutions of Equations

Many transcendental equations can be solved in the form involving W function. The solution of some of them is given below.

Equations Involving Exponential Functions

Multiplicative form: Consider the following general equation multiplicative form of exponential function:

P.K. Swamee, B.R. Chahar, *Design of Canals*, Springer Transactions in Civil and Environmental Engineering, DOI 10.1007/978-81-322-2322-1

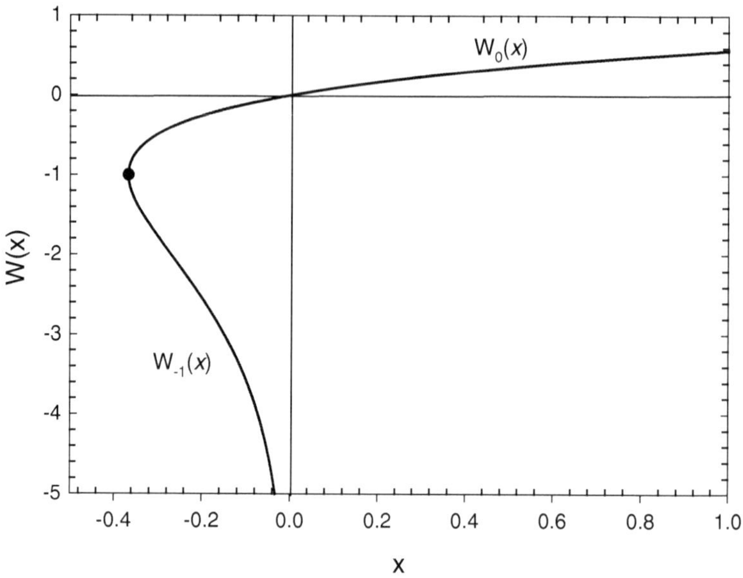

Fig. A1.1 Lambert's W function

$$x^p e^{bx^q} = c \tag{A1.2}$$

Raising *q*/*p* power on both the sides of Eq. (A1.2), one gets

$$x^q e^{bqx^q/p} = c^{q/p} \tag{A1.3}$$

Multiplying Eq. (A1.3) throughout by *bq*/*p*, the following equation is obtained:

$$\frac{bq}{p} x^q e^{bqx^q/p} = \frac{bq}{p} c^{q/p} \tag{A1.4}$$

Equation (A1.4) is of the form of Eq. (A1.1) in $(bq/p)x^q$. Thus, solution of Eq. (A1.4) is

$$\frac{bq}{p} x^q = W\left(\frac{bq}{p} c^{q/p}\right) \tag{A1.5}$$

Simplifying Eq. (A1.5), *x* is expressed as

$$x = \left[\frac{p}{bq} W\left(\frac{bq}{p} c^{q/p}\right)\right]^{\frac{1}{q}} \tag{A1.6}$$

Additive form: Consider the following form:

$$e^{bx} = x + c \tag{A1.7}$$

Equation (A1.7) is written as

$$e^{-bc} e^{b(x+c)} = x + c \tag{A1.8}$$

Equation (A1.8) is written in the following form:

$$(x + c)\, e^{-b(x+c)} = e^{-bc} \tag{A1.9}$$

Multiplying Eq. (A1.9) by $-b$,

$$-b\,(x + c)\, \mathrm{e}^{-b(x+c)} = -b\mathrm{e}^{-bc} \tag{A1.10}$$

Using Eq. (A1.1) solution of Eq. (A1.10) is

$$-b\,(x + c) = W\left(-b\mathrm{e}^{-bc}\right) \tag{A1.11}$$

Equation (A1.11) is simplified to

$$x = -c - \frac{1}{b} W\left(-b\mathrm{e}^{-bc}\right) \tag{A1.12}$$

Equations Involving Exponential Functions

Multiplicative form: Consider the following general equation involving logarithmic function:

$$x^p \ln\left(bx^q\right) = c \tag{A1.13}$$

Equation (A1.13) is expressed in the form

$$x^{-q} e^{cx^{-p}} = b \tag{A1.14}$$

Raising *p*/*q* power on both the sides of Eq. (A1.14) yields

$$x^{-p} e^{\frac{cp}{q} x^{-p}} = b^{p/q} \tag{A1.15}$$

Multiplying both the sides of Eq. (A1.15) by *cp*/*q*, one gets

$$\frac{cp}{q} x^{-p} e^{\frac{cp}{q} x^{-p}} = \frac{cp}{q} b^{p/q} \tag{A1.16}$$

Equation (A1.16) is of the form of Eq. (A1.1) in $(cp/q)x^{-p}$. Therefore, the solution of Eq. (A1.16) is

$$\frac{cp}{q}x^{-p} = W\left(\frac{cp}{q}b^{p/q}\right) \tag{A1.17}$$

Simplifying Eq. (A1.17), x is expressed as

$$x = \left[\frac{q}{cp}W\left(\frac{cp}{q}b^{p/q}\right)\right]^{\frac{1}{p}} \tag{A1.18}$$

Additive form: Consider the following form:

$$x = a\ \ln(x + c) \tag{A1.19}$$

Taking antilogarithms,

$$e^{\frac{x}{a}} = x + c \tag{A1.20}$$

Multiplying left-hand side of Eq. (A1.20) by $\exp(c/a - c/a)$,

$$\mathrm{e}^{-\frac{c}{a}}\mathrm{e}^{\frac{x+c}{a}} = x + c \tag{A1.21}$$

Dividing both the sides of Eq. (A1.21) by $1/a$,

$$\frac{1}{a}e^{-\frac{c}{a}}\mathrm{e}^{\frac{x+c}{a}} = \frac{x+c}{a} \tag{A1.22}$$

Equation (A1.22) is put to the following form:

$$-\frac{x+c}{a}e^{-\frac{x+c}{a}} = -\frac{1}{a}e^{-\frac{c}{a}} \tag{A1.23}$$

Equation (A1.23) is of the form of Eq. (A1.1). Thus,

$$x = -c - aW\left(-\frac{1}{a}e^{-\frac{c}{a}}\right) \tag{A1.24}$$

Equations Involving Power of Power Function

Consider the following general equation:

$$x^{x^p} = c \tag{A1.25}$$

Taking logarithms, Eq. (A1.25) reduces to

$$x^p \ln x = \ln c \tag{A1.26}$$

Equation (A1.25) is of the form of Eq. (A1.13). Thus, using Eq. (A1.13) and Eq. (A1.18), the solution of Eq. (A1.26) is

$$x = \left[\frac{\mathrm{W}\,(p \ln c)}{p \ln c}\right]^{-\frac{1}{p}} \tag{A1.27}$$

Simplification

Consider the following function:

$$y = x^{x^{x^{x^{\cdots}}}} \tag{A1.28}$$

Equation (A1.28) can be put in the following form:

$$y = x^y \tag{A1.29}$$

Taking logarithms and simplifying Eq. (A1.29) can be put in the following form:

$$y^{-1} \ln y = \ln x \tag{A1.30}$$

Equation (A1.30) is having the of form of Eq. (A1.13), which has got the following solution:

$$y = \frac{-\ln x}{W\,(-\ln x)} \tag{A1.31}$$

Selection of Branch of W Function

If the argument of W function is greater than or equal to zero than the principal branch, W_0 should be selected. On the other hand, on account of some physical reasoning, if the argument lies between $-1/e$ and 0, then the subsidiary branch W_{-1} is applicable. Thus, in Eq. (A1.27), the solution will lie on the principal branch W_0 if $p > 0$ and $c > 1$.

Asymptotic Limits

In 2005, for $x > 441.19$, Hassani (see Hoorfar and Hassani 2008) gave the following bounds for the W function:

$$\ln\ x - \ln\ \ln\ x \leq W_0(x) \leq \ln\ x \tag{A1.32}$$

For the magnitude of argument encountered in the normal depth problem, it is found that the lower bound given by Eq. (A1.32) is within 1 % of W_0, and the departure progressively reduces with the increase in x. Thus, one can have the following approximation for $x \geq 10^8$.

$$W_0(x) = \ln\ x - \ln\ \ln\ x \tag{A1.33}$$

Reference

Hoorfar A, Hassani M (2008) Inequalities on the Lambert W function and hyper-power function. J Inequal Pure Appl Math 9(2):1–5

Appendix 2: Conformal Mapping

The usefulness of conformal mapping in two-dimensional flow problems stems from the fact that solutions of Laplace's equation remain solutions when subjected to conformal transformation. The solution of a two-dimensional seepage problem could be reduced to one seeking the solution of Laplace's equation subject to certain boundary conditions within a region R in the z plane. Unless the region R is of a very simple shape, an analytic solution to Laplace's equation is generally very difficult. However, by means of conformal mapping, it is often possible to transform the region R into a simpler region R_1 wherein Laplace's equation can be solved subject to the transformed boundary conditions. Once the solution has been obtained in region R_1, it can be carried back by the inverse transformation to the region R of the original problem. Hence, the crux of the problem is finding a series of transformations that will map a region R conformally into a simple region R_1 such as a rectangle or a circle.

Mapping

A *complex function* is a functional relationship between two complex variables. In general, the complex functions of interest in groundwater flow are analytic. If, at each point within a region of the z plane, the function $w = f(z)$ and its derivative *dw/dz* are both single-valued and finite, then the function is said to be *analytic* within the region. An analytic function satisfies Cauchy-Riemann equations, and all the partial derivatives are continuous within the region. If a function is analytic, it possesses derivatives not only of the first order but also of all orders. Thus, the existence and continuity of all partial derivatives of real and imaginary parts of w (denoted by ϕ and ψ) are assured if $w = f(z)$ is analytic. The direct functional relationship $w = f(z)$ may not be known, but the relationship of each of the complex variables z and w with another complex variable may be known or established easily. The desired functional relationship is generally determined with the help of mapping. *Mapping* is the correspondence of a sequence of points from one plane into another plane. In groundwater, commonly used mapping functions are conformal mapping, linear mapping, inverse mapping, velocity hodograph, Zhukovsky's mapping, Schwarz-Christoffel transformation, etc. These mapping functions are described in the subsequent section following treatment used by Harr (1962).

Conformal Mapping

A transformation that possesses the property of preserving angles of intersection in magnitude and sense and the approximate image of small shapes is said to be *conformal*.

Let C (Fig. A2.1) be a smooth curve through a point z, and let C_1 be its image through point w under the transformation $w = f(z)$ when $f(z)$ is analytic at z and $f'(z) \neq 0$. As $f'(z)$ is a complex number, say $f'(z) = A \exp(i\alpha)$, then from the definition of a derivative

$$f'(z) = \frac{dw}{dz} = \lim_{\Delta z \to 0} \frac{\Delta w}{\Delta z} \tag{A2.1a}$$

the following two equations are obtained:

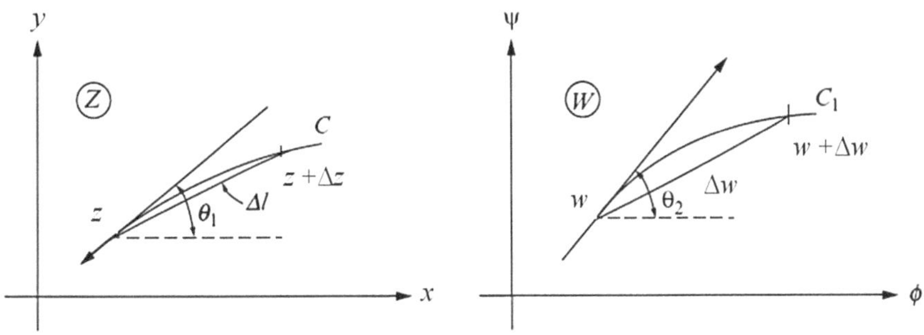

Fig. A2.1 Conformal mapping

$$\alpha = \arg\ f'(z) = \lim_{\Delta z \to 0}\ \arg \frac{\Delta w}{\Delta z} = \theta_2 - \theta_1 \tag{A2.1b}$$

$$A = \ \mathrm{mod}\ \ f'(z) = \lim_{\Delta Z \to 0} \left| \frac{\Delta w}{\Delta z} \right| \tag{A2.1c}$$

Therefore, any two curves intersecting at a particular angle at point z will, even after transformation, intersect at the same angle at w (the image of z); that is, the sides of the angle at w are rotated in the same direction by the same amount α. Infinitesimal lengths in z are magnified at w by a factor of $A = \mathrm{mod}\, f'(z)$. However, large figures in the z plane may transform into shapes bearing little resemblance to the original, but the angles formed are preserved even in these cases of intersection in case of large figures. Points at which $f'(z) = 0$ are termed as *critical* or *singular points* of the transformation where the angles are not preserved conformally. If $f'(z)$ has n-fold zeros, angles are not preserved at the critical or singular points but are multiplied $n + 1$ times.

Inverse Mapping Functions

The transformation $w = 1/z$ provides a single-valued correspondence between all points in the z and w planes; see Fig. A2.2. Taking $w = \rho \exp(i\alpha)$ and $z = r \exp(i\theta)$, the transformation yields $\alpha = -\theta$ and $\rho = 1/r$.

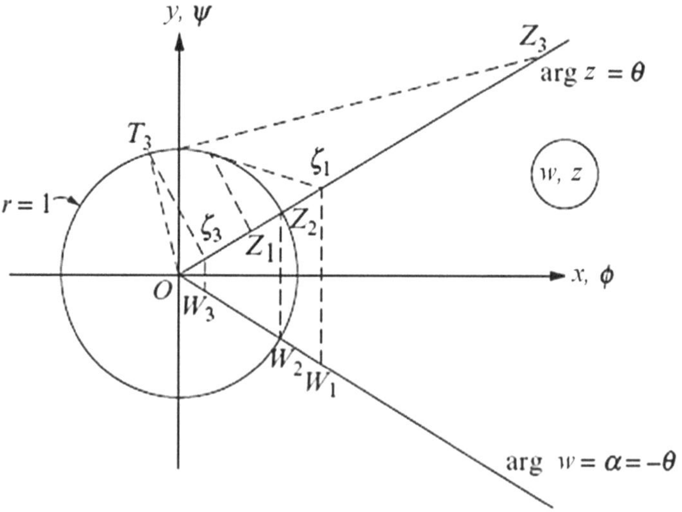

Fig. A2.2 Inverse function mapping

Thus, the transformation yields arg $w = -\arg z$ and $|w|\,|z| = 1$, which imply that there is a simple reflection about the real axis and an inversion with respect to the unit circle $r = 1$. By a familiar theorem from geometry, a point and its inverse are related as follows: As depicted in Fig. A2.2, consider a unit circle ($r = 1$) about the point O and an exterior ($r > 1$) point z_3. From z_3, construct the tangent to T_3 and drop a perpendicular to T_3 on line ζ_3. ζ_3 is said to be the inverse of point z_3, and conversely. In the inversion process, the points $r = 1$ play a special role in that they define a circle of unit radius in the z plane that maps into itself in the w plane. The points $r < 1$ in the z plane (interior points in the unit circle) correspond to the points $\rho > 1$ (exterior points in the unit circle) in the w plane, or, in general, points exterior to the unit circle are mapped into its interior. Thus, the inversion process can be thought of as a reflection about the unit circle. In particular, the point at the origin $z = 0$ or $w = 0$ is mapped into a circle at infinity $w = \infty$ or $w = \infty$. Furthermore, points at a circle of unit radius in the z (or w) plane are mapped into itself in the w (or z) plane, and points exterior to the unit circle in one plane are mapped into interior to the unit circle in the another plane and vice versa. In general, the reciprocal function transforms:

- Circles not through the origin into circles not through the origin
- Circles through the origin into straight lines not through the origin
- Straight lines through the origin into straight lines through the origin
- Straight lines not through the origin into circles through the origin

The inverse function can be used in transforming hodograph plane into inverse hodograph (*dz*/*dw*) plane. The flow region in the inverse hodograph is a polygon consisting only of straight-line boundaries.

Velocity Hodograph

Let the complex potential $w = \phi + i\psi$ be an analytical function of the complex variable z, as $w = f(z)$. Differentiation of w with respect to z yields

$$\frac{dw}{dz} = \frac{\partial \phi}{\partial x} + i\frac{\partial \psi}{\partial x} = \frac{\partial \psi}{\partial y} - i\frac{\partial \phi}{\partial y} = u - iv \tag{A2.2}$$

The transformation of the region of flow from z plane into the *dw*/*dz* plane is called the *velocity hodograph*. The utility of the hodograph stems from the fact that although the shape of the free surface and the limit of the surface of seepage are not known initially in the z plane, their hodographs are completely defined in the *dw*/*dz* plane. The elementary boundaries transform into hodograph plane as follows:

- At an impervious boundary, the velocity vector is in the direction of the boundary; in *u*-*v* plane, a straight line passing through the origin and parallel to the impervious boundary represents the impervious boundary.

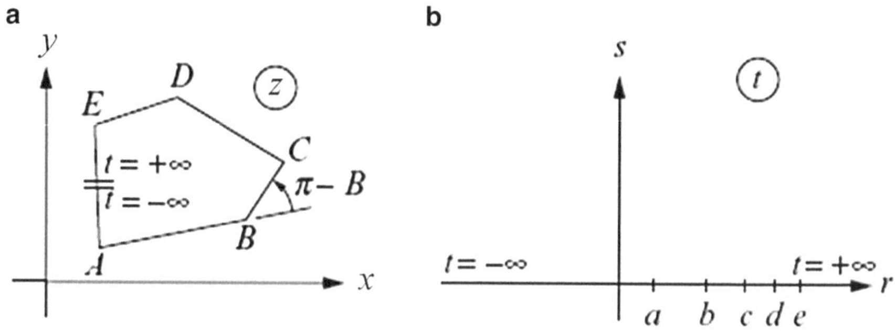

Fig. A2.3 Schwarz-Christoffel transformation

- Since a boundary of reservoir is an equipotential line, velocity vector is perpendicular to the boundary; hence, in *u*-*v* plane, a straight line passing through the origin and normal to the reservoir boundary represents the reservoir boundary.
- A line of seepage (phreatic line) is a streamline, and along it $\phi + ky = \text{const.}$; in *u*-*v* plane, a circle ($u^2 + v^2 + kv = 0$) passing through the origin with radius $k/2$ and center at $(0, -k/2)$ represents the phreatic line.
- A surface of seepage in *z* plane, which is neither a streamline nor an equipotential line, is represented by a straight line normal to the seepage surface and passing through the point $(0, -k)$ in the *u*-*v* plane.

Schwarz-Christoffel Transformation

Schwarz-Christoffel transformation is a method of mapping a polygon consisting of straight-line boundaries, from one plane onto the upper/lower half of another plane. The transformation can be considered as the mapping of a polygon from the *z* plane onto a similar polygon in the *t* plane in such a manner that the sides of the polygon in the *z* plane extend through the real axis of the *t* plane. This is accomplished by opening the polygon at some convenient point, say between A and E of Fig. A2.3a, and extending one side to $t = -\infty$ and the other to $t = +\infty$ (Fig. A2.3b). In this operation, the polygon is straightened out to a straight line extending from $t = -\infty$ to $t = +\infty$, and it is placed along the real axis of the *t* plane. See Fig. A2.3.

This transformation can be considered as the mapping of a polygon from one plane onto a similar polygon in another plane in such a manner that the sides of the polygon in one plane extend through the real axis of another plane. Thus, the transformation conformally maps the interior region of the polygon into the entire upper/lower half of another plane. The transformation that maps the *z* plane conformally onto the upper half of the ζ plane is

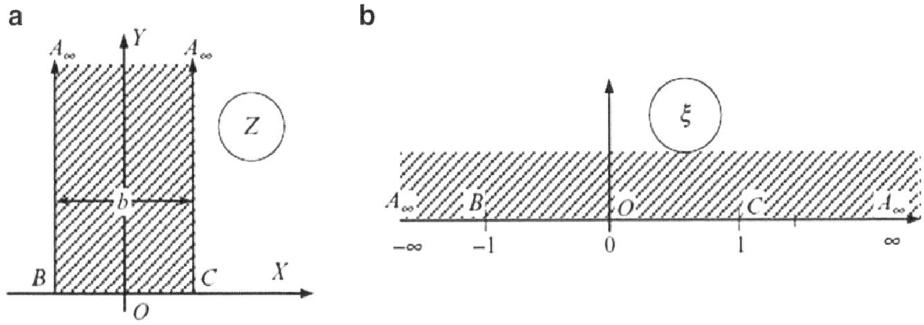

Fig. A2.4 Schwarz-Christoffel transformation of a semi-infinite strip. (**a**) Semi-infinite strip. (**b**) Upper half of auxiliary plane

$$\frac{dz}{d\zeta} = C_1(\zeta - \beta_1)^{\alpha_1 - 1}(\zeta - \beta_2)^{\alpha_2 - 1}(\zeta - \beta_3)^{\alpha_3 - 1}\cdots \quad \text{(A2.3a)}$$

$$z = C_1 \int_0^\zeta \frac{dt}{(t-\beta_1)^{1-\alpha_1}(t-\beta_2)^{1-\alpha_2}(t-\beta_3)^{1-\alpha_3}\cdots} + C_2 \quad \text{(A2.3b)}$$

where t is a dummy variable and C_1 and C_2 are complex constants. α_1, α_2, α_3, etc., are the interior angles (fraction of π) of the polygon in the w plane, and β_1, β_2, β_3, etc. ($-\infty < \beta_1 < \beta_2 < \beta_3 < \ldots < \infty$), are the points on the real axis of the ζ plane corresponding to the respective vertices. Any three of the values β_1, β_2, β_3, etc., can be chosen arbitrarily to correspond to three of the vertices of the given polygon. The remaining $N-3$ values of βs must then be determined so as to satisfy conditions of similarity. The interior angle at the point of opening may be regarded as π; hence, it takes no part in the transformation. Also the vertex of the polygon placed at infinity in the ζ plane does not appear in the transformation. Thus, by mapping a vertex of the flow region into one at infinity in the auxiliary plane omits vertex factor from the transformation and reduces one unknown.

Schwarz-Christoffel Transformation of Semi-infinite Strip

Consider the conformal mapping of the semi-infinite strip $A_\infty BC\,A_\infty$ of width b (Fig. A2.4a) onto the upper half of the ζ plane (Fig. A2.4b) through Schwarz-Christoffel transformation. The strip may be considered as a triangle with interior angles $B = \pi/2$, $C = \pi/2$, and $A = 0$. In mapping the points A_∞, B, and C arbitrarily on the points $\zeta = -\infty$, $\zeta = -1$, and $\zeta = 1$, respectively, conditions of symmetry place the fourth vertex at $\zeta = \infty$ and the Schwarz-Christoffel transformation is

$$\frac{dz}{d\zeta} = C_1(\zeta - \beta_1)^{\alpha_1 - 1}(\zeta - \beta_2)^{\alpha_2 - 1}(\zeta - \beta_3)^{\alpha_3 - 1} \quad \text{(A2.4a)}$$

$$z = C_1 \int_0^{\zeta} \frac{dt}{(t-\beta_1)^{1-\alpha_1}(t-\beta_2)^{1-\alpha_2}(t-\beta_3)^{1-\alpha_3}} + C_2 \tag{A2.4b}$$

Interior angles (fraction of π) of the polygon in the z plane are $\alpha_1 = 0$, $\alpha_2 = 1/2$, and $\alpha_3 = 1/2$; and vertices on the real axis of the ζ plane are $\beta_1 = -\infty$, $\beta_2, = -1$, and $\beta_3 = 1$. As the vertex A is placed at infinity in the ζ plane, it will not appear in the transformation. Therefore, Eq. (A2.4b) reduces to

$$z = C_1 \int_0^{\zeta} \frac{dt}{(t-1)^{1-1/2}(t+1)^{1-1/2}} + C_2 = C_1 \int_0^{\zeta} \frac{dt}{\sqrt{t^2-1}} + C_2 \tag{A2.5}$$

At the point O, $z = 0$ as well as $\zeta = 0$, so $C_2 = 0$. Using the values at point C ($z = b/2$ and $\zeta = 1$) in Eq. (A2.5),

$$\frac{b}{2} = C_1 \int_0^1 \frac{dt}{\sqrt{t^2-1}} = -iC_1 \int_0^1 \frac{dt}{\sqrt{1-t^2}} = -iC_1\left[\sin^{-1} t\right]_0^1 = -iC_1 \frac{\pi}{2} \tag{A2.6a}$$

Thus,

$$C_1 = -\frac{1}{i}\frac{b}{\pi} = i\frac{b}{\pi} \tag{A2.6b}$$

After substituting C_1 and C_2, Eq. (A2.5) becomes

$$z = i\frac{b}{\pi} \int_0^{\varsigma} \frac{dt}{\sqrt{t^2-1}} = \frac{b}{\pi}\sin^{-1}\varsigma \tag{A2.7a}$$

or

$$\varsigma = \sin\frac{\pi z}{b} \tag{A2.7b}$$

Mapping Example for Seepage from a Canal

Velocity Hodograph for Seepage from a Trapezoidal Canal

Consider a trapezoidal channel of bed width b, depth of water y, sides inclined at an angle $\pi\sigma$, or having a side slope m (1 vertical: m horizontal), underlain by a drainage layer at a depth d below the canal water surface and passing through a homogeneous

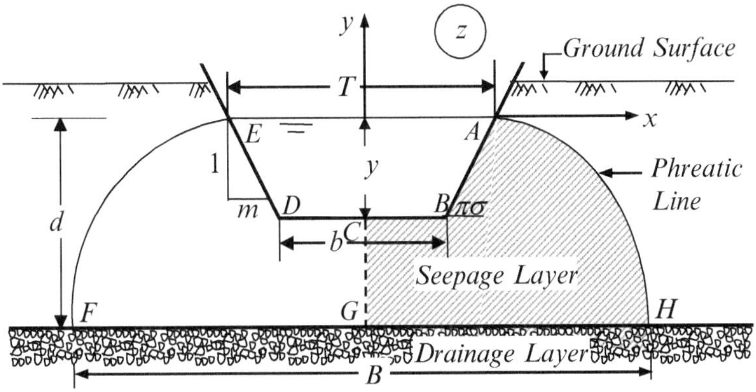

Fig. A2.5 Pattern of seepage from a trapezoidal canal with drainage layer at finite depth

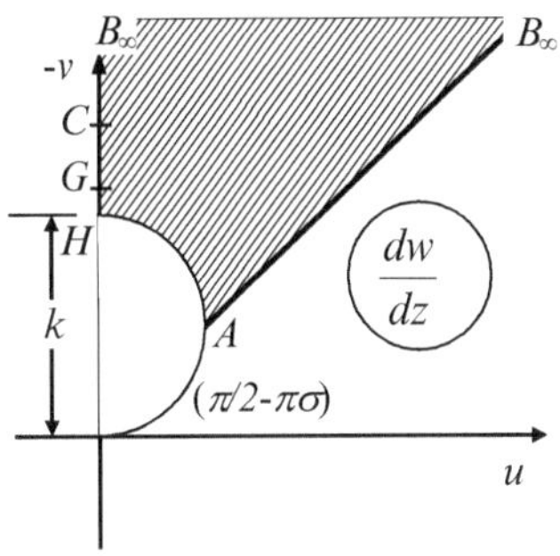

Fig. A2.6 Velocity hodograph for seepage from a trapezoidal canal

isotropic porous medium of hydraulic conductivity k as shown in Fig. A2.5 and dealt by Chahar (2007). Let the physical plane be defined as $z = x + iy$.

Let us draw the velocity hodograph plane ($dw/dz = u - iv$), for the seepage domain *ABCGHA* (shown hatched in Fig. A2.5)

It is assumed that the water table is below the top of the drainage layer, and hence, atmospheric pressure prevails at the bottom of the seepage layer and line *GH* is equipotential line. Being reservoir boundaries, *AB* and *BC* are equipotential lines as on them the piezometric head is constant. Since *AB, BC*, and *GH* are equipotential lines, so in hodograph plane, these will be straight lines passing through the origin and normal to the direction in physical seepage plane. As the point *B* is located at intersection of two reservoir boundaries with angle greater than π, it is a singular point. Thus, at point *B*, the seepage velocity is infinite. Thus, point *B* in hodograph plane goes to infinity. Point *G* maps in between *C* and *H* in hodograph plane as the seepage velocity at *H* is less than at *C*. *AH* is the phreatic line in physical plane, so it maps along a circle of radius *k* with center at (0, *k*/2) in the hodograph plane. Finally, *CG* is a flow line (like impervious boundary), so it maps a straight line passing through the origin and parallel to the direction in physical seepage plane. Therefore, the final velocity hodograph is $B_\infty CGHAB_\infty$ as shown in Fig. A2.6.

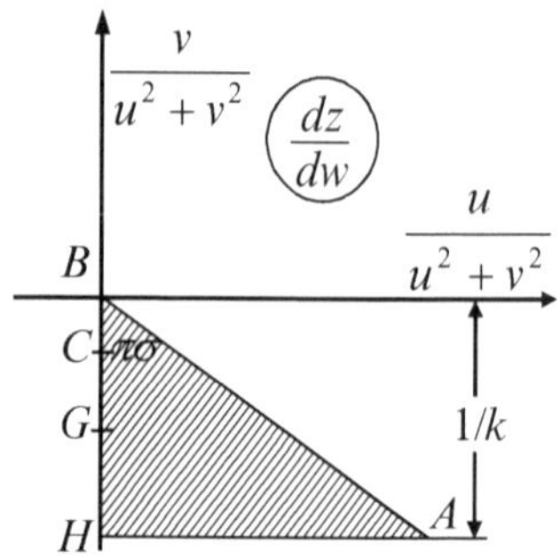

Fig. A2.7 Inverse mapping of hodograph

Inverse Hodograph Mapping

Draw an inverse hodograph plane for the case of seepage from a trapezoidal canal with drainage layer at finite depth (Fig. A2.5). It also means finding inverse mapping of velocity hodograph plane of Fig. A2.6. *AH* is a circle of diameter *k* passing through the origin in the hodograph plane, so in the inverse plane, it maps a straight line not passing through the origin and parallel to *x*-axis at distance 1/*k*. On the other hand *AB, BC, CG*, and *GH* are straight lines passing through the origin in the hodograph plane, so these lines map into straight lines through the origin in the inverse hodograph plane. Thus, the final inverse hodograph plane is *BCGHAB* as shown in Fig. A2.7.

Mapping in Complex Potential Plane

Defining complex potential $w = \phi + i\psi$ and as water table is below the top of the drainage layer and atmospheric pressure prevails at the bottom of the seepage layer, being reservoir boundary, *GH* is an equipotential line, so as *ABC*. From the conventional definition of velocity potential in porous medium, the potential difference between *ABC* and *GH* is *kd*. On the other hand, *AH* is the last flow line (phreatic line), and *CG* is the middle flow line (due to symmetry of flow in vertical plane). Thus, the difference between the flow lines *AH* and *CG* is $q_s/2$, where q_s is the seepage discharge per unit length of the canal. Therefore, the seepage domain in the physical plane is bounded by two flow lines and two equipotential lines, and hence, the final mapping in complex potential $w = \phi + i\psi$ is a rectangle *CGHA* as shown in Fig. A2.8.

Schwarz-Christoffel Transformation of Inverse Hodograph Plane

Figure A2.7 shows the inverse hodograph plane for the case of seepage from a trapezoidal canal with drainage layer at finite depth. This is accomplished by

Fig. A2.8 Complex potential mapping

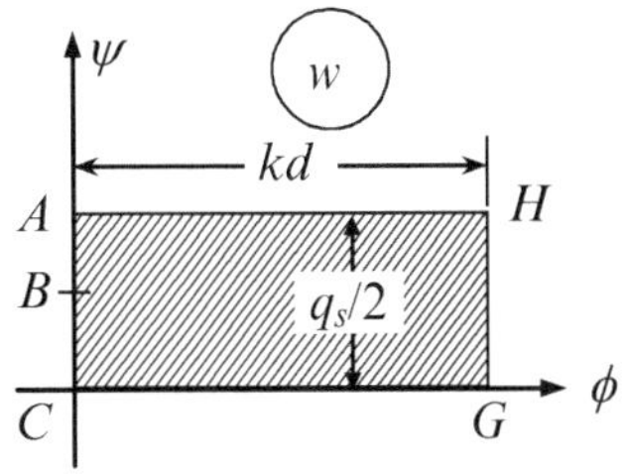

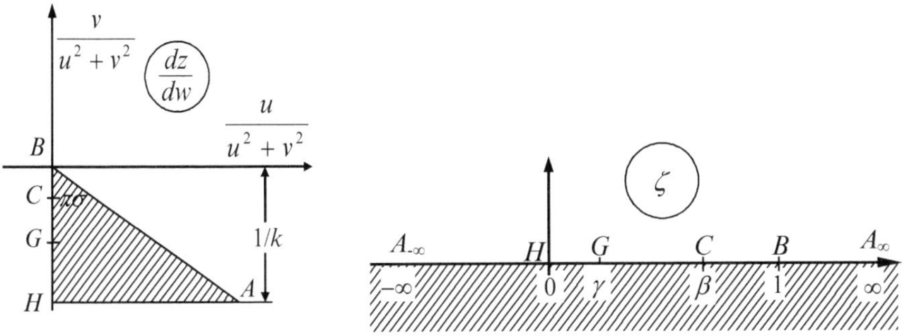

Fig. A2.9 Schwarz-Christoffel transformation of inverse hodograph plane

opening the polygon at point A and extending one side to $-\infty$ and the other to $+\infty$ and putting H at 0, G at γ, C at β, and B at 1. See Fig. A2.9.

Thus, the transformation maps conformally the region interior to the polygon of inverse hodograph into the entire lower half (i.e., maps lower half if vertices on polygon are in counterclockwise direction and upper half if they are in clockwise direction) of the auxiliary plane (ζ plane) resulting in

$$\frac{dz}{dw} = C_1 \int_0^{\zeta} \frac{dt}{(t-\beta_1)^{1-\alpha_1}(t-\beta_2)^{1-\alpha_2}(t-\beta_3)^{1-\alpha_3}(t-\beta_3)^{1-\alpha_3}} + C_2 \qquad \text{(A2.8)}$$

where interior angles of the polygon in the dz/dw plane are $\alpha_1 = 1/2, \alpha_2 = 1, \alpha_3 = 1$, and $\alpha_4 = \sigma$; and vertices on the real axis of the ζ plane are $\beta_1 = 0$, $\beta_2, = \gamma$, $\beta_3 = \beta$, and $\beta_1 = 1$. Therefore, Eq. (A2.8) reduces to

$$\frac{dz}{dw} = C_1 \int_0^{\zeta} \frac{dt}{(t-1)^{1-\sigma}\sqrt{t}} + C_2 \qquad \text{(A2.9)}$$

Constants C_1 and C_2 can be found by using values of dz/dw and ζ at two points in dz/dw plane and ζ plane. Using the values at point B ($\zeta = 1$; $dz/dw = 0$) in Eq. (A2.9),

$$0 = C_1 \int_0^1 \frac{dt}{(t-1)^{1-\sigma}\sqrt{t}} + C_2 \tag{A2.10a}$$

or

$$C_2 = -C_1 \int_0^1 \frac{dt}{(t-1)^{1-\sigma}\sqrt{t}} \tag{A2.10b}$$

Combining Eqs. (A2.9) and (A2.10b) at point H ($\zeta = 0$; $dz/dw = -i/k$),

$$-\frac{i}{k} = C_1 \int_0^0 \frac{dt}{(t-1)^{1-\sigma}\sqrt{t}} - C_1 \int_0^1 \frac{dt}{(t-1)^{1-\sigma}\sqrt{t}} \tag{A2.10c}$$

or

$$C_1 = \frac{i}{k} \Big/ \int_0^1 \frac{dt}{(t-1)^{1-\sigma}\sqrt{t}} = \frac{-ie^{-i\pi\sigma}}{k\ B\,(1/2,\ \sigma)} \tag{A2.10d}$$

Since

$$\int_0^1 \frac{dt}{(t-1)^{1-\sigma}\sqrt{t}} = \frac{1}{(-1)^{1-\sigma}} \int_0^1 (1-t)^{\sigma-1}(t)^{1/2-1}\ dt = B\,(1/2,\ \sigma)\, \mathrm{e}^{-i\pi\,(1-\sigma)} \tag{A2.11}$$

where $B(1/2,\ \sigma) =$ complete beta function, which can be expressed in terms of gamma function (Γ) as (see Abramowitz and Stegun 1972)

$$B\left(\frac{1}{2}, \sigma\right) = \frac{\Gamma\,(1/2)\,\Gamma\,(\sigma)}{\Gamma\,(0.5+\sigma)} \tag{A2.12}$$

After substituting C_1 and C_2, Eq. (A2.9) becomes

$$\frac{dz}{dw} = \frac{-ie^{i\pi\sigma}}{k\ B\,(1/2,\ \sigma)} \int_1^{\varsigma} \frac{dt}{(t-1)^{1-\sigma}\sqrt{t}} \tag{A2.13}$$

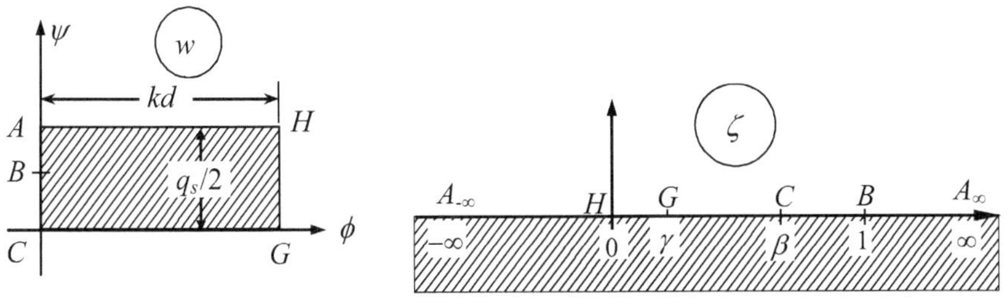

Fig. A2.10 Schwarz-Christoffel transformation of complex potential plane

Quantity of Seepage

In the w plane, the mapping is a rectangle as shown in Fig. A2.10, which can be mapped onto a similar polygon in auxiliary plane by opening the polygon at A point and extending one side to $-\infty$ and the other to $+\infty$ and putting H at 0, G at γ, C at β, and B at 1 as shown in Fig. A2.10.

Thus, the transformation maps conformally the region interior of w plane into the entire lower half of the auxiliary plane (ζ plane) resulting in

$$w = C_3 \int_0^{\zeta} \frac{dt}{(t-\beta_1)^{1-\alpha_1}(t-\beta_2)^{1-\alpha_2}(t-\beta_3)^{1-\alpha_3}(t-\beta_3)^{1-\alpha_3}} + C_4 \tag{A2.14}$$

where C_3 and C_4 are complex constants; interior angles (fraction of π) of the polygon in the w plane are $\alpha_1 = ½$, $\alpha_2 = ½$, $\alpha_3 = ½$, and $\alpha_4 = 1$; and vertices on the real axis of the ζ plane are $\beta_1 = 0$, $\beta_2, = \gamma$, $\beta_3 = \beta$, and $\beta_1 = 1$. Therefore, Eq. (A2.14) reduces to

$$w = C_3 \int_0^{\zeta} \frac{dt}{\sqrt{t(t-\gamma)(t-\beta)}} + C_4 \tag{A2.15}$$

The constants C_3 and C_4 have been determined using the values at points C ($\zeta = \beta$; $w = 0$) and G ($\zeta = \gamma$; $w = kd$). After substituting C_3 and C_4, Eq. (A2.15) becomes

$$w = \frac{kd}{\int_{\beta}^{\gamma} \frac{dt}{\sqrt{t(t-\gamma)(t-\beta)}}} \int_{\beta}^{\zeta} \frac{dt}{\sqrt{t(t-\gamma)(t-\beta)}} \tag{A2.16}$$

Further, it is known that

$$\int_{\beta}^{\gamma} \frac{dt}{\sqrt{t(t-\gamma)(t-\beta)}} = -i\int_{\gamma}^{\beta} \frac{dt}{\sqrt{t(t-\gamma)(\beta-t)}} = -\frac{2i}{\sqrt{\beta}}K\left(\sqrt{(\beta-\gamma)/\beta}\right) \tag{A2.17}$$

where $K\left(\sqrt{(\beta-\gamma)/\beta}\right)$ = complete elliptical integral of the first kind with modulus $\sqrt{(\beta-\gamma)/\beta}$ (Byrd and Friedman 1971). Therefore,

$$w = i\frac{kd\sqrt{\beta}}{2K\left(\sqrt{(\beta-\gamma)/\beta}\right)}\int_{\beta}^{\zeta} \frac{dt}{\sqrt{t(t-\gamma)(t-\beta)}} \tag{A2.18}$$

Differentiating Eq. (A2.18) with respect to ζ,

$$\frac{dw}{d\zeta} = i\frac{kd\sqrt{\beta}}{2K\left(\sqrt{(\beta-\gamma)/\beta}\right)}\frac{1}{\sqrt{\zeta(\zeta-\gamma)(\zeta-\beta)}} \tag{A2.19}$$

Since $\frac{dz}{d\zeta} = \frac{dz}{dw}\frac{dw}{d\zeta}$, substitution of $\frac{dz}{dw}$ from Eq. (A2.13) and $\frac{dw}{d\zeta}$ from Eq. (A2.19) results in

$$\frac{dz}{d\zeta} = \frac{e^{i\pi\sigma}}{B(1/2,\sigma)}\frac{d\sqrt{\beta}}{2K\left(\sqrt{(\beta-\gamma)/\beta}\right)}\frac{1}{\sqrt{\zeta(\zeta-\beta)(\zeta-\gamma)}}\int_{1}^{\zeta} \frac{dt}{(t-1)^{1-\sigma}\sqrt{t}} \tag{A2.20}$$

Integrating Eq. (A2.20) and applying the condition at H ($\zeta = 0$; $z = B/2 - id$) gives

$$z = \frac{B}{2} - id + \frac{e^{i\pi\sigma}}{B(1/2,\sigma)}\frac{d\sqrt{\beta}}{2K\left(\sqrt{(\beta-\gamma)/\beta}\right)}\int_{0}^{\zeta}\left(\int_{1}^{t} \frac{d\tau}{(\tau-1)^{1-\sigma}\sqrt{\tau}}\right) \times\frac{dt}{\sqrt{t(t-\beta)(t-\gamma)}} \tag{A2.21}$$

where τ = another dummy variable. Equation (A2.21) defines the physical domain of the seepage flow $ABCGHA$. Using Eq. (A2.21) at G ($\zeta = \gamma$; $z = -id$); C ($\zeta = \beta$; $z = -iy$); B($\zeta = 1$; $z = b/2 - iy$); and A ($\zeta = \infty$; $z = b/2 + y\cot\pi\sigma$) and thereafter manipulation yield

$$\frac{B}{d} = \frac{\sqrt{\beta}}{K\left(\sqrt{(\beta-\gamma)/\beta}\right)\ B\,(1/2,\sigma)}\int_0^{\gamma}\frac{F_1\,(t,\ \sigma)\,dt}{\sqrt{t\,(\beta-t)\,(\gamma-t)}} \tag{A2.22a}$$

$$\frac{d}{y} = \frac{K\left(\sqrt{(\beta-\gamma)/\beta}\right)\ B\,(1/2,\sigma)\,/\,\sqrt{\beta}}{K\left(\sqrt{(\beta-\gamma)/\beta}\right)\ B\,(1/2,\sigma)\,/\,\sqrt{\beta}-\frac{1}{2}\int_{\gamma}^{\beta}\frac{F_1\,(t,\sigma)\,dt}{\sqrt{t\,(\beta-t)\,(t-\gamma)}}} \tag{A2.22b}$$

$$\frac{b}{y} = \frac{\int_{\beta}^{1}\frac{F_1\,(t,\sigma)\ dt}{\sqrt{t\,(t-\beta)\,(t-\gamma)}}}{K\left(\sqrt{(\beta-\gamma)/\beta}\right)\ B\,(1/2,\sigma)\Big/\sqrt{\beta}-\frac{1}{2}\int_{\gamma}^{\beta}\frac{F_1\,(t,\sigma)\ dt}{\sqrt{t\,(\beta-t)\,(t-\gamma)}}} \tag{A2.22c}$$

where

$$F_1\,(t,\sigma) = \int_t^1\frac{dt}{(1-\tau)^{1-\sigma}\sqrt{\tau}} = B\,(1/2,\sigma) - B_t\,(1/2,\sigma) \tag{A2.22d}$$

and $B_t(1/2,\,\sigma)$ = incomplete beta function (Abramowitz and Stegun 1972). Using the values at point H ($\zeta = 0$; $w = kd + iq_s/2$) in Eq. (A2.18),

$$kd + \frac{q_s}{2}i = \frac{i\ kd\,\sqrt{\beta}}{2K\left(\sqrt{(\beta-\gamma)/\beta}\right)}\int_{\beta}^{0}\frac{dt}{\sqrt{t\,(t-\gamma)\,(t-\beta)}}$$

$$= \frac{kd\,\sqrt{\beta}\int_0^{\beta}\frac{dt}{\sqrt{t\,(t-\gamma)\,(\beta-t)}}}{2K\left(\sqrt{(\beta-\gamma)/\beta}\right)} \tag{A2.23}$$

Separating real and imaginary parts leads to

$$q_s = 2kd\frac{K\left(\sqrt{\gamma/\beta}\right)}{K\left(\sqrt{(\beta-\gamma)/\beta}\right)} \tag{A2.24}$$

Therefore, the seepage expression becomes

$$F_s = 2\frac{d}{y}\frac{K\left(\sqrt{\gamma/\beta}\right)}{K\left(\sqrt{(\beta-\gamma)/\beta}\right)} \tag{A2.25}$$

This involves two transformation parameters β and γ. Simultaneous solution of Eqs. (A2.22b) and (A2.22c) for a given channel dimension (b, y, and σ) and depth of the drainage layer d results in parameters β and γ. Using these values in Eq. (A2.25), the seepage function and then the quantity of seepage is determined. Chahar (2007) includes detailed solution for seepage from a trapezoidal canal with drainage layer at finite depth.

References

Abramowitz M, Stegun IA (1972) Handbook of mathematical functions with formulas, graphs, and mathematical tables. Dover Pub., New York

Byrd PF, Friedman MD (1971) Handbook of elliptic integrals for engineers and scientists. Springer, Berlin

Chahar BR (2007) Analysis of seepage from polygon channels. J Hydraul Eng ASCE 133(4):451–460

Harr ME (1962) Groundwater and seepage. McGraw-Hill, New York

Appendix 3: Solution of Cubic Equation

Without loss of generality, a cubic equation is written as

$$x^3 + ax^2 + bx + c = 0 \tag{A3.1}$$

Taking $x = y + k$, Eq. (A3.1) changes to

$$y^3 + (3k + a)\,y^2 + \left(3k^2 + 2ak + b\right)y + k^3 + ak^2 + bk + c = 0 \tag{A3.2}$$

Putting

$$k = -a/3 \tag{A3.3}$$

Equation (A3.1) changes to

$$y^3 + Ay + B = 0 \tag{A3.4}$$

where

$$A = b - \frac{a^2}{3} \tag{A3.5}$$

$$B = c + \frac{2a^3}{27} - \frac{ab}{3} \tag{A3.6}$$

One Real Root Case

Substituting $y = p + q$, Eq. (A3.4) takes the form

$$p^3 + q^3 + (3pq + A)(p + q) + B = 0 \tag{A3.7}$$

Adopting

$$pq = -A/3, \tag{A3.8}$$

Eq. (A3.7) reduces to

$$p^3 + q^3 = -B \tag{A3.9}$$

Cubing Eq. (A3.8), one gets

$$p^3q^3 = -A^3/27 \tag{A3.10}$$

Eliminating q between Eqs. (A3.9) and (A3.10),

$$p^6 + Bp^3 - A^3/27 = 0 \tag{A3.11}$$

Solving Eq. (A3.11) as a quadratic in p^3, p is found to be

$$p = \left(-\frac{B}{2} \pm \sqrt{\frac{B^2}{4} + \frac{A^3}{27}}\right)^{\frac{1}{3}} \tag{A3.12}$$

Similarly, q is obtained as

$$q = \left(-\frac{B}{2} \mp \sqrt{\frac{B^2}{4} + \frac{A^3}{27}}\right)^{\frac{1}{3}} \tag{A3.13}$$

Adding p and q from Eqs. (A3.12) and (A3.13), y is obtained as

$$y = \left(-\frac{B}{2} + \sqrt{\frac{B^2}{4} + \frac{A^3}{27}}\right)^{\frac{1}{3}} + \left(-\frac{B}{2} - \sqrt{\frac{B^2}{4} + \frac{A^3}{27}}\right)^{\frac{1}{3}} \tag{A3.14}$$

As there are three cube roots of unity 1, ω, and ω^2, the other two roots of Eq. (A3.4) are

$$y = \left(-\frac{B}{2} + \sqrt{\frac{B^2}{4} + \frac{A^3}{27}}\right)^{\frac{1}{3}} \omega + \left(\frac{B}{2} - \sqrt{\frac{B^2}{4} + \frac{A^3}{27}}\right)^{\frac{1}{3}} \omega^2 \tag{A3.15}$$

$$y = \left(-\frac{B}{2} + \sqrt{\frac{B^2}{4} + \frac{A^3}{27}}\right)^{\frac{1}{3}} \omega^2 + \left(-\frac{B}{2} - \sqrt{\frac{B^2}{4} + \frac{A^3}{27}}\right)^{\frac{1}{3}} \omega \qquad \text{(A3.16)}$$

Knowing y, x is obtained as $x = y - a/3$. It can be seen from Eqs. (A3.14), (A3.15), and (A3.16) that for

$$\frac{B^2}{4} + \frac{A^3}{27} \geq 0 \qquad \text{(A3.17)}$$

there is one real root and two complex conjugate roots. On the other hand, for

$$\frac{B^2}{4} + \frac{A^3}{27} < 0 \qquad \text{(A3.18)}$$

there are three real roots expressed in a combination of complex terms, which is not convenient to compute.

Three Real Roots Case

The following trigonometric identity

$$\cos\ 3\theta = 4\cos^3\theta - 3\ \cos\theta \qquad \text{(A3.19)}$$

is written as

$$\cos^3\theta - \frac{3}{4}\cos\theta - \frac{1}{4}\cos\ 3\theta = 0 \qquad \text{(A3.20)}$$

Suppose

$$y = \lambda\cos\theta \qquad \text{(A3.21)}$$

Putting Eq. (A3.21) in Eq. (A3.4), and rearranging, one gets

$$\cos^3\theta + \frac{A}{\lambda^2}\cos\theta + \frac{B}{\lambda^3} = 0 \qquad \text{(A3.22)}$$

Assuming $A/\lambda^2 = -3/4$, one finds λ as

$$\lambda = 2\sqrt{-\frac{A}{3}} \qquad \text{(A3.23)}$$

Substituting λ from Eq. (A3.23) to Eq. (A3.22),

$$\cos^3\theta - \frac{3}{4}\cos\theta - \frac{1}{4}\sqrt{-\frac{27B^2}{4A^3}} = 0 \tag{A3.24}$$

Comparing Eqs. (A3.20) and (A3.24), one gets

$$\cos\ 3\theta = \pm\sqrt{-\frac{27B^2}{4A^3}} \tag{A3.25}$$

Using Eqs. (A3.21) and (A3.23), one gets the first root of Eq. (A3.4) as

$$y_1 = 2\sqrt{-\frac{A}{3}}\cos\theta \tag{A3.26}$$

The other two roots of Eq. (A3.4) are

$$y_2 = 2\sqrt{-\frac{A}{3}}\cos\left(\theta + \frac{2\pi}{3}\right) \tag{A3.27}$$

$$y_3 = 2\sqrt{-\frac{A}{3}}\cos\left(\theta + \frac{4\pi}{3}\right) \tag{A3.28}$$

Knowing y, x is obtained as $x = y - a/3$.

Example 1 Solve $x^3 + 20x^2 + 10x + 10 = 0$.

Solution For $a = 20$, $b = 10$, and $c = 10$, $A = -123.333$ and $B = 535.926$, so $\frac{B^2}{4} + \frac{A^3}{27} = 2321.284 > 0$ indicates one real root of the equation. Equation (A3.14) yields $y = -12.847$, and hence, $x = -19.514$. Substituting x in the given cubic equation, one gets 0.0005.

Example 2 Solve $x^3 + 20x^2 + 10x - 10 = 0$.

Solution For $a = 20$, $b = 10$, and $c = -10$, $A = -123.333$ and $B = 515.926$, so $\frac{B^2}{4} + \frac{A^3}{27} = -2937.975 < 0$ indicates that there are three real roots of the given equation. Considering lower sign in Eq. (A3.25), $\cos(3\theta) = -0.9786$ yielding $3\theta = 2.9345$. Therefore, Eqs. (A3.26), (A3.27), and (A3.28) give $y_1 = 7.1626$, $y_2 = 5.6304$, and $y_3 = -12.7930$, and hence, $x_1 = 0.4959$, $x_2 = -1.0362$, and $x_3 = -19.4597$. Substituting x_1, x_2, and x_3 in the given cubic equation, the following values are obtained: 0.0000, 0.0002, and 0.0004.

Index

P.K. Swamee, B.R. Chahar, *Design of Canals*, Springer Transactions in Civil and Environmental Engineering, DOI 10.1007/978-81-322-2322-1